SF₆
switchgear

Other volumes in this series

Volume 1 Power circuit breaker theory and design
C. H. Flurscheim (Editor)
Volume 2 Electric fuses
A. Wright and P. G. Newbery
Volume 3 Z-transform electromagnetic transient analysis in high-voltage networks
W. Derek Humpage
Volume 4 Industrial microwave heating
A. C. Metaxas and R. J. Meredith
Volume 5 Power system economics
T. W. Berrie
Volume 6 High voltage direct current transmission
J. Arrillaga
Volume 7 Insulators for high voltages
J. S. T. Looms
Volume 8 Variable frequency AC motor drive systems
D. Finney
Volume 9 Electricity distribution network design
E. Lakervi and E. J. Holmes

SF6 switchgear

HM Ryan & GR Jones

Peter Peregrinus Ltd. on behalf of the Institution of Electrical Engineers

Published by: Peter Peregrinus Ltd., London, United Kingdom

© 1989: Peter Peregrinus Ltd.

All rights reserved. No part of this publication may be reproduced, stored in a retrieval system or transmitted in any form or by any means—electronic, mechanical, photocopying, recording or otherwise—without the prior written permission of the publisher.

While the author and the publishers believe that the information and guidance given in this work are correct, all parties must rely upon their own skill and judgment when making use of them. Neither the author nor the publishers assume any liability to anyone for any loss or damage caused by any error or omission in the work, whether such error or omission is the result of negligence or any other cause. Any and all such liability is disclaimed.

British Library Cataloguing in Publication Data

Ryan, H.M.
   SF_6 switchgear
   1. Electrical equipment. Switchgear & controlgear—Manuals
   I. Title. II. Jones, G.R. (Gordon Rees) 1938-
III. Series
621.31'7
ISBN 0 86341 123 1

Printed in England by Biddles Ltd., Guildford

Contents

		Page
1	Introduction	1
2	Fundamental properties of SF_6	6
3	Types of SF_6 Interrupters	15
4	Characteristics of SF_6 Interrupters	22
4.1	Pressure characteristics	22
4.2	Thermal characteristics	31
4.2.1	Gas heating	31
4.2.2	Nozzle ablation	33
4.3	Electromagnetic characteristics	37
4.4	Thermal-recovery characteristics	43
4.4.1	Axisymmetric (puffer-type) interrupters	43
4.4.1.1	Nozzle and contact considerations	44
4.4.1.2	Piston-travel and contact-throttling effects	50
4.4.1.3	SF_6 Mixtures	51
4.4.1.4	Liquid SF_6 injection	53
4.4.2	Asymmetric (rotary arc) interrupters	53
4.5	Dielectric recovery	54
4.5.1	Dielectric recovery of the remnant arc column	55
4.5.2	Scaling of the dielectric recovery of the arc column	58
4.5.3	Retarded dielectric recovery	60
4.5.4	Dielectric-recovery requirements for medium-voltage interrupters	61
4.5.5	Gases with higher dielectric strength	62
5	Arc modelling and computer-aided methods for interrupter-design evaluation	63
5.1	Field calculations	64
5.2	Arc modelling	69
5.3	Puffer modelling	72
5.4	Rotary-arc interrupter modelling	73

vi Contents

6	**Impact of SF_6 technology upon transmission switchgear**		**76**
6.1	Commercial considerations		80
6.2	Circuit-breaker assemblies		82
6.3	Metalclad installations		84
	6.3.1	Components and structure	88
	6.3.2	Insulation co-ordination	95
	6.3.3	Internal arcing faults in metalclad enclosures	100
	6.3.4	Internal maintenance requirements and reliability	103
6.4	Artificial current zeros		103
	6.4.1	Generator circuit breakers	104
	6.4.2	High-voltage DC circuit breakers	107
6.5	Particular examples of SF_6-insulated installations		113
7	**Impact of SF_6 technology upon distribution and utility switchgear**		**116**
7.1	Operation and system requirements		116
7.2	Relative merits of SF_6, vacuum and more traditional circuit breakers		118
7.3	Puffer circuit breakers		121
7.4	Rotary-arc circuit breakers		122
7.5	SF_6 self-extinguishing circuit breakers		124
7.6	Insulation of distribution switchgear		126
7.7	Fuse–switch combinations		127
7.8	Disconnecting and earthing switches		132
8	**Operating mechanisms for SF_6 circuit breakers**		**134**
8.1	Energy requirements		134
8.2	Reliability		136
8.3	Puffer circuit-breaker mechanisms		136
8.4	Choice of drive type for puffer interrupters		139
8.5	Modelling puffer-drive mechanisms		140
9	**Impact of SF_6 technology upon regulations, testing and instrumentation**		**143**
9.1	Circuit-breaker testing		143
	9.1.1	Electrical tests	143
		9.1.1.1 Test circuits	144
		9.1.1.2 Recent modifications to the basic Weil Circuit	147
		9.1.1.3 Test methods for three phases in the tank breaker	149
		9.1.1.4 Unit testing of multibreak tank-type circuit breakers	151
		9.1.1.5 Synthetic tests for closing and auto-reclosing duties	154
		9.1.1.6 Short-circuit tests for disconnecting switches	156
		9.1.1.7 Synthetic tests for high-voltage DC circuit breakers	158
	9.1.2	Mechanical tests	160
	9.1.3	Chemical tests	162
	9.1.4	Particular performance capabilities of SF_6 circuit breakers	165

	9.2	Instrumentation and diagnostics	167
	9.2.1	Electrical measurements	167
	9.2.2	Mechanical-drive measurements	170
	9.2.3	Aerodynamic measurements	172
	9.2.4	Radiation measurements	173
	9.2.5	Chemical measurements	176

10 Conclusions   **178**

11 References   **180**

     Index   **192**

Acknowledgments

The authors wish to acknowledge the assistance provided by their many colleagues and acquaintances worldwide. Their willingness and prompt responses contributed to the acquisition of updated information. The unstinting effort of Miss E. Kevan with the typescript is also greatly appreciated.

Acknowledgment is made to the following organisations for the use of illustrations: Brown Boveri Co. Ltd.; Brush Switchgear Ltd.; GE Co. USA; GEC High Voltage Switchgear Ltd.; A/S Norsk Elektrisk; Bonneville Power Administration, Oregon, USA; ASEA High Voltage Switchgear; Siemens AG; Mitsubishi Electric Corporation; NEI Reyrolle Ltd.; Hitachi Ltd.; Toshiba Corporation; South Wales Switchgear Ltd.; Merlin–Gerin; McGraw-Edison Co.; Kansai Electric Power Co. Inc.; Sprecher Energy; Tokyo Electric Power Co. Inc.; Ontario Hydro; Westinghouse R&D Center; Electric Power Development Co. Ltd.; Yaskawa Electric Manufacturing Co. Ltd.; Yorkshire Switchgear Group; Square D Co.; KEMA High Power Laboratories; Fuji Electric Corporation R&D Ltd.; CESI; Oak Ridge Nuclear Laboratories, USA.

Chapter 1

Introduction

The purpose of a circuit breaker is to ensure the unimpeded flow of current in a network under normal operating conditions, and to interrupt the flow of excessive current in a faulty network. It may also be required to interrupt load current under some circumstances and to perform an open–close–open sequence (auto-reclosing) on a fault in others. The successful achievement of these duties relies upon the availability of good mechanical design to meet the demands of opening and closing the circuit-breaker contacts, and good electrical design to ensure that the circuit breaker can satisfy the electrical stresses.

During the opening and closing sequences an electric arc occurs between the contacts of the circuit breaker, and advantage is taken of this discharge to assist in the circuit-interruption process. For instance, in an AC network, the arc is tolerated in a controlled manner until a natural current zero of the waveform occurs when the discharge is rapidly quenched to limit the reaction of the system to the interruption. With asymmetric waveforms and for DC interruption, advantage is taken of the arc resistance for damping purposes or to generate a controlled circuit instability to produce an artificial current zero. The arc control demanded by such procedures may require gas pressurisation and flow, which in turn make additional demands upon the operating mechanism.

Although this description is simplified it serves to illustrate the complexity of the interactions involved in circuit interruption. These interactions are determined on the one hand by the nature of the arcing and arc quenching medium, and on the other by the network demands. Since Garrard's review (1976), increasing use has been made of SF_6 as the arc-quenching medium, initially perhaps because of its outstanding electrical-insulation properties and its chemical inertness under normal conditions. However, with increased usage there was a growing awareness that SF_6 also possessed attractive arc quenching properties in its own right. Subsequently, there was a realisation that it also possessed compressive and thermal absorption properties, which are sufficiently different from those of other interrupter media such as oil and air, so that different modes of utilisation in an interrupter environment could be used to advantage. These have led to the evolution of puffer, suction, self-pressurising and rotary-arc

2 Introduction

interrupters which have been commercially successful on account of their cost effectiveness and reduced size. The latter feature along with the excellent insulation strength of SF_6 has led to the evolution of metalclad substation systems within which the circuit breaker forms a single unit. As a consequence, the size, design, installation and appearance of such substations have been revolutionised.

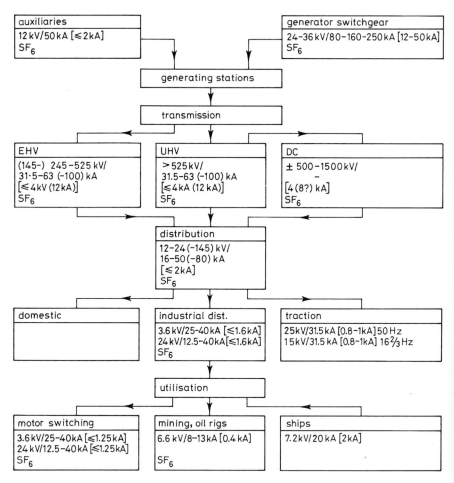

Fig. 1.1 *Range of switchgear applications: Voltage, fault current [normal current]*

The range of applications of SF_6 switchgear is extensive and is mainly governed by network demands. The switchgear rating associated with each application can only be typically specified on the understanding that invariably exceptions are likely to exist. If this latter limitation is accepted a classification of applications with typical ratings may be assembled as shown in Fig. 1.1. Since a major aspect of power switchgear is protection, not only is it important to

Introduction   3

know what is being protected and what level of protection is provided, but also it is helpful to be aware of the other links in the overall protection chain since the specified local protection assumes that the preceding link carries its own adequate protection. (Occasionally some trade-offs may be possible between adjacent links or within a link, e.g. compromise between switchgear and fuse rating.) For this reason, the classification of applications is presented sequentially from generation through transmission and distribution to utilisation (Fig. 1.1). The ratings are given in terms of voltage, fault current and normal current (square brackets).

Although the most common transmission is at EHV, both UHV and DC transmission systems have been commissioned since Garrard's review (1976). Secondary transmission is still mainly at 145–245 kV and distribution at 12–24 kV, although 145 kV may still be used by Area Boards for distribution in heavily loaded systems (Garrard, 1976).

In the case of high-voltage AC transmission, the growth in power transmission has been, and is projected for the future to be, accompanied by increases in transmission voltages (Fig. 1.2.*a*) and rated currents (Fig. 1.2*b*). Furthermore, the rising generating capacities of power stations and the increasing density of networks are together producing even higher short-circuit current levels (Fig. 1.2*c*). Transmission-switchgear capabilities have needed to keep pace with the increased protection demands being made by such developments.

In order to place the impact being made by SF_6 switchgear in perspective it is worth recalling the comments made by Garrard (1976) in his review. At that time, the main challenge to the traditional types of circuit breakers (oil and air) came from vacuum circuit breakers in the lower range of distribution voltage, and from SF_6 at transmission voltages. It was expected that both vacuum and SF_6 would capture a larger share of the market, but that there would be a middle voltage and current range in which the low-oil circuit breakers would maintain an important position. Totally enclosed metalclad SF_6 insulated gear was already competing strongly with open-type substations, and it was expected that British manufacturers would have a complete range from 145 to 420 kV available by 1980. Circuit breakers for high-voltage DC transmission were at an early stage of development and synthetic test methods had become an economic necessity.

Since then, SF_6 circuit breakers have indeed made a substantial impact upon the switchgear market, not only at extra-high-voltage levels but now increasingly in the lower-voltage ranges used for distribution and utilisation. This impact exceeds that envisaged by Garrard (1976), in that SF_6 circuit breakers are available for almost all the applications listed in Fig. 1.1 including generator and high-voltage DC protection duties. More recently, there are indications that SF_6-based switches are being considered for voltages below 3·6 kV. Even the expendiency of using synthetic test methods is currently being extended to its limits by the need to undertake simultaneous testing of three phases at voltages up to 145 kV. For the higher transmission voltages where multibreak-per-pole

4 Introduction

circuit breakers are used, recourse is being made to the unit testing of each pole on account of the limitations of synthetic power sources.

This book places an emphasis upon research and development and the use of computer-aided-design methods, since it is clear that we are currently in the

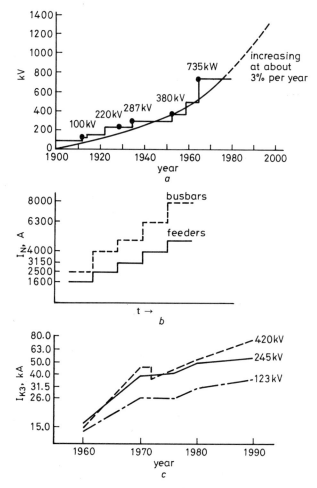

Fig. 1.2   Ratings of AC power systems (Figs. 1–3, Elmcke, 1981)
 a Transmission voltages
 b Rated currents
 c Maximum short-circuit currents

middle of an evolutionary cycle as far as SF_6 interrupters are concerned, which draws increasingly from such modern technological methods. Indeed the importance of such computer-aided-design methods is recognised by CIGRE (Study Comittee 13) through their establishment of a working Group on Circuit-Breaker Arc Modelling.

Developments in the field of high-voltage-transmission protection and in the protection of distribution systems are considered separately on account of the possible use of different interrupter principles in the associated switchgear. The impact of SF_6 technology upon IEC standards, circuit-breaker testing and measuring instrumentation is also considered.

Chapter 2
Fundamental Properties of SF_6

Pure SF_6 gas is colourless, odourless, tasteless and non-toxic, it is chemically stable and non-flammable. At room temperature and pressure it is gaseous, and at $6.616 \, gl^{-1}$ it has a mass density of 4.7 times that of air. Since its critical temperature is 45.6°C (Fig. 2.1a) SF_6 can be liquefied by compression, and therefore may be conveniently stored as a liquid. On the other hand, liquefaction in a circuit breaker must be avoided since the change of phase from gas to liquid leads to a rapid loss of gas density and a possible reduction in insulation strength. The minimum temperature commonly specified for outdoor electrical equipment is −25°C which implies that the maximum gas pressure at 20°C which can be used without heaters is 1.9 bar (gauge) (Fig. 2.1b). However, this limitation is offset by the superior dielectric strength of SF_6. At a given pressure, SF_6 has a dielectric strength several times higher than that of air, and at a pressure of 1.9 bar has a dielectric strength equivalent to that of switch oil (Fig. 2.1b).

Of course, the choice of SF_6 for use in high-power circuit breakers does not rely solely upon its good dielectric strength, but it depends also upon its excellent arc-quenching and control properties. In order to understand the diverse ways in which arc quenching is achieved a knowledge of the properties of SF_6 at elevated temperatures up to about 20 000 K is required. At such temperatures SF_6 dissociates into a large number of fragments (Fig. 2.1c), so that the gas, which is inert at room temperature, becomes composed of highly chemically reactive ions. Consequently careful choice of materials for the construction of the arcing chamber is required as is the need for high SF_6 purity (and, in particular, the absence of hydrogen which might lead to the formation of HF). The time scale on which some of these fragments recombine may be relatively long, so that the equilibrium concentrations given in Fig. 2.1c are modified as shown in Fig. 2.1d. Many of these persisting fragments are toxic and are the subject of many investigations (e.g. Sauers *et al.*, 1984). However, it is the formation of these chemically reactive and toxic fragments that contributes to the good arc control, quenching and dielectric recovery which can be achieved with SF_6.

For instance, the hot dielectric-recovery period following arcing will be governed by the temperatures and persisting species shown in Fig. 2.1d. The ease with which negative ions are formed owing to the affinity for free electrons leads to the superior dielectric-recovery properties of SF_6. Furthermore the dissociation of SF_6 into these fragments at a temperature of about 2000 K (Fig. 2.1d) leads to peaks in the thermal conductivity of about an order of magnitude greater than ambient (Fig. 2.1e). During the thermal-recovery period at the end of an arcing half cycle it is thermal radial diffusion (as approximately represented by the thermal conduction) at these temperatures which is the controlling mechanism on account of the small arc-column cross-section and steep peripheral temperature profiles. Thus the dissociation of SF_6 into these fragments in the temperature range just below the temperature for the onset of significant electrical conductivity (Fig. 2.1f) is one factor which leads to a remarkably good and rapid thermal recovery (of the order of 3 μs compared with 8 μs for air).

Not only does SF_6 have such superior properties for ensuring a better and more rapid recovery of dieletric strength following arcing, but it also possesses a range of properties which are particularly attractive for arc control during the high-current arcing phase and for the possibility of utilising a number of fundamentally different methods for arc control. Thus its electrical conductivity is comparable with copper vapour at temperatures above 8000 K (Fig. 2.1f). Of course, at lower temperatures it has a substantially lower conductivity so that contamination with copper vapour is to be avoided in order to preserve thermal and dielectric recovery. These good electrical-conduction properties at high temperatures reduce advantageously the power dissipated within the circuit breaker by a factor of about 20% compared with air at currents of a few tens of kiloamperes.

A major power-dissipating process (although not the only one) from such high-current arcs is by radiative transfer. This can be quantified in terms of a net emission coefficient (Fig. 2.1g). Part of this radiation escapes totally from the arc environment whilst another part is absorbed by the gas surrounding the arc column. The part escaping totally is ultimately absorbed by the walls of the interrupter chamber, producing under extreme conditions ablation of the wall material. The part of the radiation absorbed by the surrounding gas heats the gas to an intermediate temperature. The enormous scale of the radiated power from high-current circuit-breaker arcs is set by the fact that a 100 kA arc in air contaminated by a high prcentage of copper vapour emits 10 MW to the wall of the interrupter chamber and so may cause considerable damage.

The net radiative power from an SF_6 arc is less than that from an air arc carrying the same current. A greater proportion of the emitted radiation is absorbed by the surrounding SF_6 gas on account of its more complex molecular structure and formation of dissociation fragments. Consequently less radiation damage occurs to the circuit-breaker enclosure with SF_6 than with air, and a more controlled heating of the surrounding gas can be achieved. This would be modified for both air and SF_6 if copper contamination with its high radiative

8 Fundamental properties of SF_6

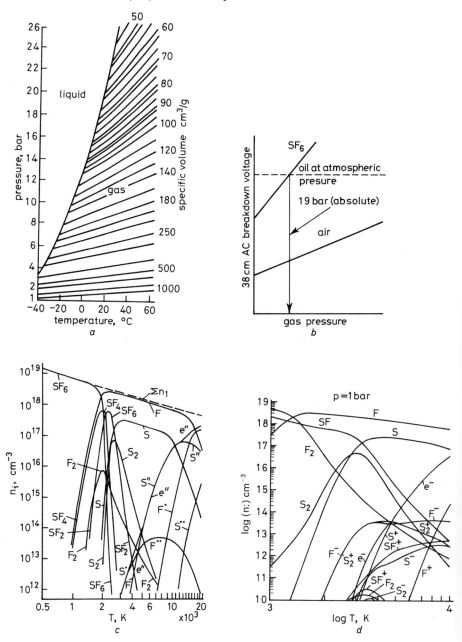

Fig. 2.1 *Properties of SF_6*
 a Pressure: temperature diagram for SF_6 (Fig. 1, Stewart, 1979)
 b Dielectric strength of SF_6, oil and air (Fig. 2, Stewart, 1979)
 c Total particle densities in SF_6 (Fig. 5, Kopainsky, 1978. Copyright Plenum Press)
 d Partial equilibrium densities for 1 bar (Fig. 6, Kopainsky, 1978. Copyright Plenum Press)

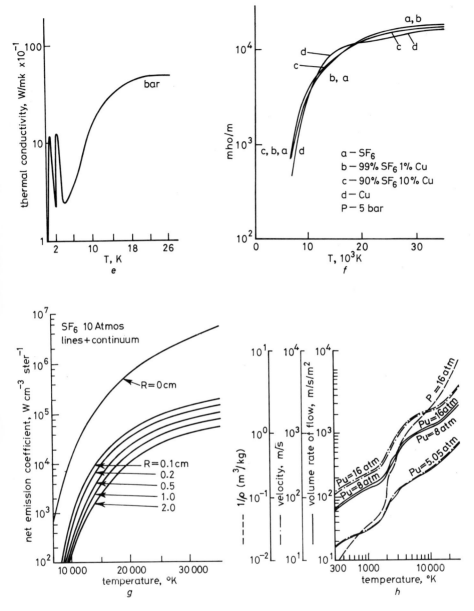

Fig. 2.1 (continued)
e Thermal-conductivity: temperature (Frost and Libermann, 1971. Copyright IEEE)
f Electrical-conductivity: temperature (Fig. A3, Airey, 1976. Copyright IEEE)
g Calculated net emission coefficient for various arc-column radii at 10 atm (Fig. 1, Libermann and Lowke, 1976)
h Specific volume, sonic velocity and volume rate of flow for SF_6 (Fig. 3, Ueda et al., 1979)

10 Fundamental properties of SF_6

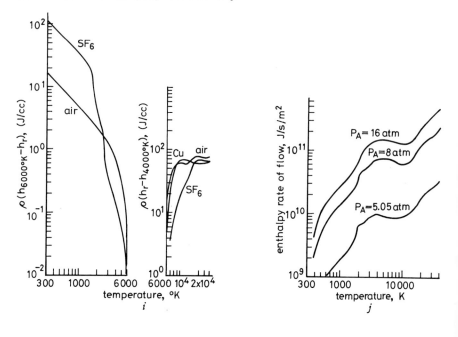

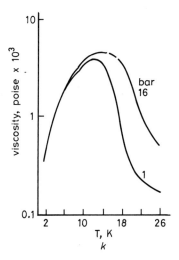

Fig. 2.1 *(continued)*

 i Mass-density: enthalpy-difference product (Fig. 7, Ueda et al., 1979)
 j Enthalpy-rate-of-flow: temperature at various pressures (Fig. 2, Ueda et al., 1979)
 k Viscosity-coefficient: temperature (Frost and Libermann, 1971. Copyright IEEE)

Fundamental properties of SF_6   11

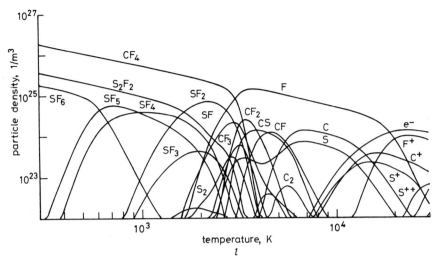

Fig. 2.1 *(continued)*
l Particle densities for a 1:1 mixture of SF_6-PTFE (Fig. 9, Ruchti, 1985)

emission were present. Nonetheless this aspect of radiative emission, although not yet fully understood, is particularly important since it has, in conjunction with some additional unique thermal and aerodynamic properties of SF_6, led to the evolution of new forms of SF_6 interrupters which would not be commercially feasible with air.

The ability of a gas to absorb thermal energy is determined by the product of the mass density and enthalpy changes (i.e. thermal capacity). Both these properties vary with gas temperature, but it is only the mass density which varies substantially with pressure (Fig. 2.1h).

The product of mass density and enthalpy change is shown for SF_6, air and copper vapour (Fig. 2.1i) for the two temperature ranges – room temperature to 6000 K, and 6000 K to 40 000 K. For temperatures less than about 2500 K SF_6 has a better ability to absorb heat than air, and indeed SF_6 at a temperature of 1500 K has the same absorption ability as air at room temperature. Thus heated SF_6 is as efficient for absorbing heat as room-temperature air. For temperatures above 6000 K the heat content of SF_6 is less than both air and copper vapour, which implies that, during arc-column recovery, less heat needs to be dissipated from the SF_6 plasma than from either air or copper plasma.

In Fig. 2.1h it is the reciprocal of mass density ($1/\varrho$) which is shown. This parameter is known as the specific volume and, of course, increases substantially with temperature. The implication of this variation is that the volume of a mass of SF_6 gas increases by about 10^3 with an increase in temperature from room temperature up to 20 000 K at constant pressure. Of course at constant volume this would produce a considerable pressure increase, the magnitude of which would be determined by the thermal-energy input and the thermal capacity.

12 Fundamental properties of SF_6

Thus the lower the heat capacity of the gas, the greater the temperature and pressure generated. But because of its considerably higher heat capacity at the lower temperatures, SF_6 is able to absorb significantly more thermal energy than air and so avoid excessive and destructive pressure changes which would otherwise occur with air. This has important safety implications regarding arcing in SF_6-filled metalclad systems and important operational implications for arc-induced pressurisation in SF_6 interrupters. The tolerance and utilisation of such an induced pressurisation could not have been accepted with so much confidence with air compared to SF_6.

If there is no mixing of heated and cool gas in a closed volume, pressurisation still occurs on account of the displacement of mass from the growing heated volume into the decreasing cooler volume. With SF_6 this can lead to pressure increases which are approximately double those which would occur if the hot and cool gases were completely mixed (Ueda et al. 1979).

In SF_6 interrupters which utilise convection for arc control and interruption, the pressure rise due to arc heating will be moderated by the flow of gas through a nozzle. For pressures above critical, the volume flow rate is governed by the specific volume and sonic velocity, all of which increase substantially with temperature (Fig. 2.1h). The results of Fig. 2.1h show that the volume flow rate at room temperature is about 30 times as large as that at 20 000 K. Since this ratio is much smaller than that for the change in specific volume discussed above ($\times 10^3$), the increase in volume of arc-heated gas cannot be compensated for by the exhausted volume if the latter is at such an elevated temperature. Consequently, even with hot-gas exhaust, substantial pressurisation can still be achieved through arc heating.

The effectiveness of convection for removing thermal energy is governed by the specific-enthalpy flow rate (Fig. 2.1j), which is the product of mass density, specific enthalpy and sound volocity. Thus, although the mass flow rate at room temperature is less than at higher temperatures, the specific-enthalpy flow rate is greater at the higher temperatures. Furthermore, since the electrical conductivity of SF_6 is negligibly small for temperatures below 3000 K (e.g. Frost and Liebermann, 1971), and since therefore there is no significant electrical power input at these temperatures, it is feasible to utilise arc-heated SF_6 within this temperature range for enthalpy removal. Compared with air, SF_6 offers a number of significant advantages in this respect. Within this temperature range, the sound velocity in SF_6 increases more rapidly than for air due to the dissociation of the SF_6 molecule (Fig. 2.1h), and becomes approximately the same as for air. Furthermore, the specific enthalpy of SF_6 is greater than air up to 2500 K. These factors, coupled with the negligible electrical power input, make SF_6 superior for enthalpy removal at these elevated temperatures than air.

There are also additional benefits to be derived from these unique thermodynamic properties of SF_6. Since the improved specific-enthalpy flux is gained substantially from the considerable change in enthalpy rather than a corresponding change in mass density, the mass of SF_6 available for enthalpy

Fundamental properties of SF_6 13

removal is more efficiently preserved than with air. This makes the use of single-pressure interrupters (puffer, self pressurising, rotary arc) a considerably more attractive possibility with SF_6. Furthermore, any limitation imposed by a relatively low nominal pressure of 5·9 bar, to avoid liquefaction at $-25°C$, becomes even less of a restriction owing to the possibility of generating higher pressures at higher temperatures due to arc heating during operation.

Another aerodynamic property which has generally received less attention in relation to switchgear arcs is the viscosity. The viscosity coefficient of SF_6 increases by an order of magnitude in the temperature range 2000–10 000 K (Fig. 2.1k), before decaying to a value similar to that at ambient conditions at 20 000 K. The implication of this result is that momentum transfer across the boundary of the arc column is restricted and leads to vortex formation and turbulence in the case of axisymmetric flow arcs (Niemeyer and Ragaller, 1973), and drag effects associated with internal circulating flows in the case of crossflow arcs (e.g. Jones and Fang, 1980). The precise influence of such effects upon circuit-breaker arc performance warrants further study, although the Brown Boveri Group (e.g. Hermann et al., 1976) have made significant progress in understanding such effects.

The above discussion concerns the properties of pure SF_6, whereas in practice a number of interrupter situations may occur when significant contamination of the SF_6 prevails. Although the material properties of these contaminated systems may in principle be evaluated, the calculations are often hampered by the lack of atomic data, ignorance of the mixture composition or non-equilibrium effects. Probably the three contamination systems most relevant to SF_6 interrupters are SF_6–nitrogen (which are of interest for overcoming liquefaction problems in low-temperature applications), SF_6–Cu vapour (which arises from the entrainment of contact vapour into the SF_6 plasma) and SF_6–PTFE mixtures (which arise from the ablation of the circuit-breaker nozzle).

In the case of N_2–SF_6 mixtures some indication of property variations is given from an inspection of the properties of the two pure species, the nitrogen values not being significantly different from the air values already discussed above (e.g. Fig. 2.1i).

The major influences of copper vapour in the arc plasma are twofold. First, the copper vapour enhances radiation loss (e.g. Strachan et al., 1977), so that the net emission coefficient (Fig. 2.1g) is increased above the SF_6 values. Secondly, the electric conductivity of the plasma is modified as shown in Fig. 2.1f). Significantly, at lower temperatures than those shown in Fig. 2.1f the coppervapour conductivity becomes greater than that of pure SF_6, although at the higher temperatures ($> 20 000 K$) the opposite is true.

The situation for SF_6–PTFE mixtures is more complicated in that account needs to be taken of the dissociation of the PTFE as well as the SF_6. Thus the equilibrium composition data of Fig. 2.1c is modified, for example, to that shown in Fig. 2.1l for a 1:1 mixture of SF_6 and PTFE (Ruchti, 1985). The presence of additional fluorine from the SF_6 prevents the formation of graphite at temperatures below about 4000 K. The presence of a more stable CF_4 component in the plasma mixture shifts the onset of dissociation from about 1500 K (Fig. 2.1c) for pure SF_6 to about 2500 K for SF_6–PTFE. The electrical conductivity of the SF_6–PTFE mixture is similar to that of SF_6 (Fig. 2.1f) for temperatures above 6000 K, whereas below this temperature SF_6 has a slightly higher conductivity. These comments apply only on the assumption that thermal equilibrium prevails, and may not always be valid in real circuit-breaker situations.

Further consideration of SF_6 mixtures is given in Section 4.4.1.3.

Chapter 3

Types of SF_6 interrupters

The use of air as an arc-quenching medium in EHV interrupters necessitated the generation of a supersonic flow coaxial with the arc and produced by suitably designed nozzles. The objective was to prevent excessive blocking of the flow during the peak current phase (when the arc cross-section could approach that of the nozzle), so that the arc plasma and surrounding hot gas could be efficiently removed before the critical recovery period following the current zero at the end of the current half-cycle. The flow in such a system was sustained by gas stored at a sufficiently high pressure to ensure a strong supersonic flow. A sufficient mass of gas was needed to ensure interruption should the fault persist and more air needed to be compressed following completion of the interrupting operation. The use of such a 2-pressure system becomes redundant if air is replaced by SF_6, because of the limited pressure which may be used and because of the more suitable compressive properties of SF_6.

For many applications two-pressure air systems have been replaced by SF_6 puffer systems, which overcome the need to store large quantities of high-pressure gas for prolonged periods. Instead, pressurisation to produce sufficient flow is achieved by piston compression during contact separation. At higher current levels, arc heating is utilised to 'block' the nozzle in a controlled manner (as indicated in Chapter 2), in order to preserve valuable arc-quenching gas until the appropriate recovery period. There are four distinct types of puffer interrupters (Fig. 3.1), various manufacturers preferring each different form. The simplest is the monoflow system utilising unidirectional flow from the piston chamber. This type of puffer concept has been superseded by more sophisticated forms designed, amongst other factors, to reduce contact vapour and its deleterious effects upon performance.

The other forms of puffer interrupters utilise the duoflow principle, whereby the compressed gas may escape in opposing directions. In the partial duo-blast interrupter (Fig. 3.1b), the main flow is through a large-diameter, normally insulating, nozzle, whilst subsidiary flows occur through both contacts, which are hollow. These subsidiary contraflows not only limit the entrainment of contact material into the main insulated nozzle, but also probably assist in

16 Types of SF$_6$ interrupters

providing a gentler interruption at lower fault currents when the contact separation is insufficient to expose the main flow through the large nozzle. This form of interrupter is favoured amongst others by NEI (Reyrolle) Ltd.

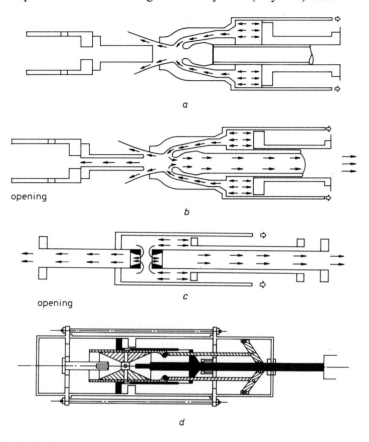

Fig. 3.1 Puffer interrupters
a Mono-blast
b Partial duo-blast (Fig. 1, Ali and Headley, 1984)
c Duo-blast (conducting nozzles)
d Duo-blast (insulating nozzles) (Fig. 1.6, Noeske et al., 1983)

The full duo-blast puffer (Fig. 3.1c) utilises two matched, large nozzles which also serve as the contacts of the interrupter. Consequently, for this form of puffer the nozzles are conductors. Nonetheless metal-vapour entrainment is limited by the bidirectional flow. This may be further reduced through the use of graphite for the construction of the downstream nozzle section, where the arc roots are eventually displaced by the flow. This type of interrupter may be regarded as two monoflow breaks in tandem but with two of the upstream contacts advantageously eliminated. This design is favoured by GEC (UK) and Siemens.

Types of SF_6 interrupters 17

An attempt has been made by GE (US) to design a full duo-blas interrupter with insulating nozzles, so that full advantage can be taken of optimised down-stream flow produced by detailed profiling of the nozzle and without interference by arc-root movement (Fig. 3.1*d*). In this design the contacts are solid and contact vapour entrainment is again limited by the two opposing flows. The authors are not aware that this design is in commercial production.

Unfortunately, piston-driven puffer circuit breakers make severe demands on the piston drive mechanism in order to overcome nozzle blocking and arc-induced pressurisation. Thus although the puffer principle is successfully utilised by many manufacturers for distribution-system applications (e.g. Brush Switchgear, Yorkshire Switchgear, NEI (Reyrolle) Ltd., GEC Switchgear) the quest for a means to overcome this limitation has seen a number of developments, particularly at the lower-voltage levels. Takahashi *et al.* (1974) have proposed using the fault current itself to electromagnetically drive the piston (Fig. 3.2*a*).

Piston-induced compression may in principle be totally eliminated and reliance placed upon the arc-induced pressurisation to provide the required arc-quenching flow at current zero (Fig. 3.2*b*). The use of this so-called 'self pressurising' principle is attractive on account of the large energies which are provided by the arc at high fault-current levels. On the other hand, at low fault-current levels the pressurisation is limited because of insufficient arc heating, and the gas-conserving advantages of nozzle blocking are not manifest on account of too small an arc cross-section. It is therefore necessary to rely upon 'free burning arcing' (no constraining forced convection) to meet the performance-voltage-restrike specifications, which are by IEC standards most severe at such low fault currents. However, provided prolonged arcing is permitted at such lower power levels, it may be possible to meet the specification simply by sufficient arc lengthening. The method has been investigated by Toshiba Corporation for use in earthing switches.

A further evolution of the puffer principle, with the objective of achieving the

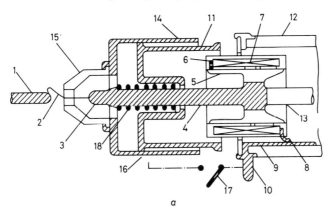

Fig. 3.2 *Self-pressurising interrupters*
a Magnetically driven piston (Fig. 6, Takahashi et al., 1974. Copyright IEEE)

18   Types of SF$_6$ interrupters

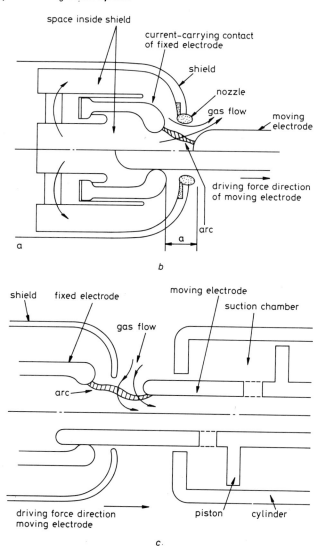

Fig. 3.2   *(continued)*
   b Arc heated pressurisation (Fig. 4a, Suzuki et al., 1984)
   c Suction type (Fig. 4c, Suzuki et al., 1984)

required performance with reduced drive energy, involves using an auxiliary piston to generate flow through gas suction outside the main arc chamber (Fig. 3.2c). Unlike the puffer case, the suction piston does not act against the arc-induced pressurisation, and is therefore less demanding of energy to drive the piston so that the advantages of self-pressurisation are maintained. This principle has been investigated by Mitsubishi Electric Corporation.

   Alternative methods for providing high-current arc control and current-zero

Types of SF_6 interrupters

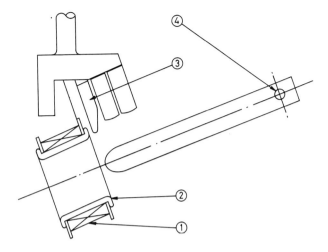

1 field coil
2 arcing tube
3 fixed arcing contact
4 contact bar pivot

a

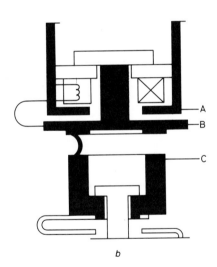

b

Fig. 3.3 *Rotary-arc interrupters*
a *Helical arc (12 kV) (Fig. 1, Parry, 1984)*
b *Ring contacts (Fig. 17, Duplay and Hennebert, 1983. Merlin–Gerin)*

20 Types of SF_6 interrupters

arc quenching for distribution applications involve rotating the arc electromagnetically (rotary-arc interrupters) (Fig. 3.3). The driving Lorentz force is produced by the fault current through the arc and by the magnetic field produced by passing the fault current also through a solenoid. Thus, unlike the puffer-type interrupters which utilise superimposed axisymmetric flows, the rotary-arc interrupters are non-axisymmetric and the arc, rather than the surrounding gas, suffers most movement to generate an effective cross-flow. Two distinct types of geometries are commercially available. The first (Fig. 3.3*b*) involves rotating the arc along two annular contacts and is preferred by Merlin–Gerin. The second geometry (Fig. 3.3*a*) involves rotating the arc azimuthally between a rod contact and annular contact which also serves as a yoke to support the magnetic-field-producing coil. This geometry is preferred by South Wales Switchgear and Brush Switchgear. It has the advantage of forming the arc in a region of maximal magnetic-flux density and encouraging a helical penetration of the arc into a coil. On the other hand, the Merlin–Gerin design does not necessarily involve arc transfer following contact separation. Both types of interrupter are, however, subjected to low-fault-current limitations like the self-pressurising interrupters, since the electromagnetic driving force decreases with fault current. Increasing the number of coil turns to overcome this difficulty is limited by considerations of mechanical strength to withstand the excessive magnetic forces which would be produced at the higher values of fault current. Nonetheless, these difficulties have been overcome by optimisation, and successful interrupters are now commercially available.

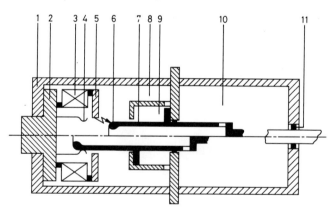

Fig. 3.4 *Hybrid interrupters*
  a Brown Boveri type (Fig. 2, Jakob et al., 1985. Copyright Brown Boveri)

Hybrid interrupters, using self-pressurisation along with the rotary-arc principle for more efficient arc-induced heating and pressurisation, are also available. For instance, Brown Boveri produced such a hybrid interrupter with low fault-current interruption enhanced by auxilary piston action (Fig. 3.4*a*).

In general, the puffer-type interrupter is attractive for medium- and high-voltage system applications, and less attractive for lower-voltage applications on account of the driving energy and hence size of mechanism required. The self-pressurising breaker is becoming increasingly attractive for medium-voltage applications on account of the reduced demands upon mechanical drive. The rotary-arc interrupter is finding increased use for low- to medium-voltage levels since no piston drive mechanism is required, so permitting the use of simple contact-separation mechanism.

Chapter 4

Characteristics of SF$_6$ interrupters

The characteristics of SF$_6$ interrupters are governed by the particular manner in which transient pressure elevation is utilised in the puffer-type interrupter, and electromagnetic arc rotation in the case of the rotary-arc interrupters. Considerable effort is being made to understand how interrupter performance may be optimised through a control of these phenomena. We therefore attempt to summarise the general conclusions emerging from these investigations, being nonetheless conscious that, because of the complexity and close coupling of the many phenomena involved, it is difficult, if not impossible, to derive simple universally applicable design guidance.

4.1 Pressure characteristics

Pressure transients produced during arcing may arise due to a number of effects (e.g. Jones, 1984), all of which are associated with the formation of the heated (but electrically nonconducting) gas surrounding the arc column. Since, as already indicated in Chapter 2, the mass density in this heated volume is considerably less than ambient (Fig. 2.1h), its formation implies the exclusion of a substantial mass of gas. Thus, within the confines of a flow-shaping nozzle, the existence of such a heated volume reduces the effective flow cross-section with a consequent modification of the pressure gradient. If, in addition, the rate of growth of the thermal volume exceeds the mean cold gas-flow velocity through the nozzle (which is likely at high peak currents), then an additional pressure elevation occurs, owing to the accumulation of mass within the nozzle approaches, which cannot be transported through the nozzle. This impulsive behaviour can lead to the excitation of aerodynamic resonances within the interrupter, manifest by relatively high-frequency pressure oscillations and which may be strongly coupled to instabilities of the arc column itself (Leclerc *et al.*, 1980). Should the thermal volume be sufficiently large to penetrate into the upstream plenum, then clearly more extensive pressure elevations may occur on account of the larger volume involved. Simultaneously, some of the aerodynamic resonances may be attenuated due to parts of the system becoming overdamped. If, in addition, substantial mixing of the heated and cool gas in the upstream plenum occurs, then the pressure elevation may be substantially modified and approximately determined by the simple gas law. This situation occurs when rotary arc motion is utilised on account of the increased swirl and turbulence promoting better mixing. Finally, the optical radiation from the arc column (Chapter 2) which is not absorbed by the heated surrounding gas ($\sim$ 10–35% of the total radiation) is absorbed by the nozzle wall. Under overload conditions, the wall material ablates, leading to the injection of additional mass into the nozzle volume and making additional demands upon the mass-transport capabilities of the system. Thus under extreme conditions, when the mass ablation rate exceeds the mass throughput of the nozzle, a reverse flow into the upstream plenum occurs, so producing a further pressure elevation. In commercial SF_6 interrupters, all these pressure-elevation-inducing processes may occur, and one aspect of a successful design evolution is to take full advantage of the more beneficial properties of the most dominant process.

In order to highlight the more sophisticated use which is made in SF_6 puffer interrupters of induced pressure transient, some pressure records from a 2-pressure and a commercial puffer interrupter are shown in Fig. 4.1a and b (Shimmin et al., 1985). Both interrupters, operated in the same current range, utilised similar nozzle geometries, but the puffer interrupter was of the partial duoflow type (Fig. 3.1b), whereas the 2-pressure interrupter employed monoflow. The operational philosophy for a 2-pressure breaker is clear, whereby excessive pressure due to nozzle blocking [and even backflow at 54 kA (Fig. 4.1a)] is tolerated provided the pressure has been reduced to nominal by the recovery phase at the end of the half-cycle of current, as an indication of a sufficiently early relief from nozzle blocking. On the other hand, in the case of the puffer interrupter, the pressure during the recovery phase remains considerably elevated above the no-load pressure (Fig. 4.1b), partly on account of the piston chamber gas being conserved due to simple nozzle blocking but also to varying extents due to the other arc-induced effects described above. For instance, excitation of aerodynamic resonances is apparent on the test records under load conditions, but not in the no load test records. Also noteworthy is the decompression which is apparent on the two pressure test result at the end of a current half-cycle, and which may be due to a combination of flow inertia and arc-heating effects. The most significant differences in the pressure characteristics of 2- and single-(puffer) pressure interrupters are summarised in Fig. 4.1c (Shimmin et al., 1985). These results show how the puffer-breaker design successfully promotes a higher pressure during the important arc-recovery period than at current peak, and how this recovery-period pressure increases more strongly with peak current than does the corresponding pressure at current peak. By contrast, in the 2-pressure breaker, the recovery-period pressure is somewhat below nominal, whilst the peak current pressure is so excessive as to produce undesirable levels of backflow.

24   Characteristics of SF_6 interrupters

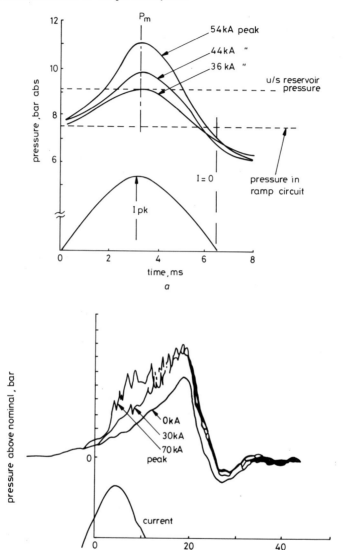

Fig. 4.1 *Arc induced pressure transients*
a 2-pressure breaker (Fig. 2a, Shimmin et al., 1985)
b Puffer breaker (Fig. 2b, Shimmin et al., 1985)

Much work has been performed worldwide to examine the manner in which gas properties, geometry and operating conditions combine to produce a net pressure rise (e.g. Natsui *et al.*, 1977; Tominaga *et al.*, 1979; Calvino *et al.*, 1976; Noeske, 1977; Yoshioka *et al.*, 1979; Yoshioka and Nakagawa, 1978, 1980; Murai *et al.*, 1981; Ueda *et al.*, 1981; Ushio *et al.*, 1981; Sasao *et al.*, 1982) in

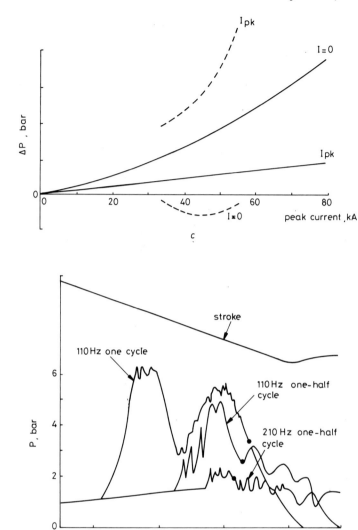

Fig. 4.1 *(continued)*
  c Comparison of 2-pressure (– – –) and puffer (———) breaker transients (Fig. 3, Shimmin et al., *1985)*
  d Influence of number of current loops and frequency on puffer pressure transients (Figs. 6 and 16, Yoshioka and Nakagawa, 1980. Copyright IEEE)

a puffer-type interrupter so that a more unified understanding is beginning to emerge.

The influences of the fault-current waveform, the point-on-current-wave contact separation, the point-on-piston compression arcing and the speed of

26 Characteristics of SF_6 interrupters

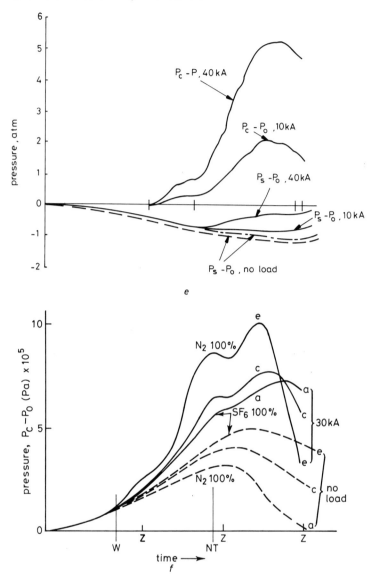

Fig. 4.1 *(continued)*
e Suction breaker (Fig. 4, Murai et al., 1981. Copyright IEEE)
f Puffer with gas mixtures (Fig. 7, Sasao et al., 1982. Copyright IEEE)

contact movement have been investigated in a series of experiments performed by Yoshioka and Nakagawa (1978, 1980).

Fig. 4.1d is a typical example of a series of pressure measurements made by using a model puffer at reduced current levels (< 10 kA) and higher waveform frequencies (110 and 210 Hz). Regardless of whether the puffer operation is

Characteristics of SF_6 interrupters 27

optimised, and in contrast to conditions which prevail in the commercial EHV circuit breaker (Fig. 4.1b), the recovery-period pressure is less than the peak current pressure, probably on account of disproportionally reduced piston compression and piston-chamber volume. There is evidence, however, that this second form of pressure transient may be commercially favoured for distribution-switchgear applications, on account of economic savings arising from a less powerful drive and a physically smaller device.

Thus, for this second type of operating regime, the investigations of Yoshioka and Nakagawa (1978, 1980) allow a number of practically useful conclusions to be drawn. By utilising a low-level steady holding current of variable duration before the main current half-cycle the magnitude of the arc-induced pressure rise may be determined as a function of the instant during the piston and contact stroke at which the current half-cycle is initiated. In this way, the optimum instant during the stroke for producing the maximum pressure rise may be determined. Although not explicitly stated, this optimum time corresponds to the instant just prior to the nozzle being unblocked as it is drawn away from the solid contact. This suggests that the highest pressure is produced by prolonging the blockage of the nozzle during the stroke through the use of the arc to replace the contact, whilst simultaneously achieving the maximum arc heating in the piston chamber with the longest length of arc up to the throat of the nozzle.

The influence of current waveform upon the pressure transient is apparent from Fig. 4.1d, the results again being obtained using a low-level steady holding current. For this type of operating regime the pressure elevations during each of the two half-cycles and full-cycle current waveform are similar and there is no substantial accumulation of pressure from the first half-cycle. The single half-cycle of arcing, delayed until contact separation has brought into action the nozzle, produces a similar pressure transient to the full-cycle case although the current-zero pressure is somewhat depressed in the half-cycle case. The use of the delayed half-cycle arcing, for initial design studies concerning tolerable pressure levels, would therefore appear to be feasible, at least in some circumstances.

Detailed information concerning the influence of asymmetric current waveforms upon pressure elevation is limited, although Yoshioka *et al.* (1979) make the obvious observation that the recovery-period pressure is less following the minor current loop. The low-value steady holding-current waveform followed by a full half-cycle of current, could possibly provide reasonable simulation of a minor followed by a major loop.

The influence of geometry upon puffer-pressure elevation can only be meaningfully considered by taking simultaneous account of the nozzle geometry, the piston chamber and the contact size and movement, since the influences of all three factors are closely coupled in the puffer-type interrupter.

Yoshioka and Nakagawa (1980) have investigated the influence of a moving solid contact in throttling the on-load flow through a nozzle. The piston-chamber volume, contact diameter and contact travel speed remained fixed, the different throttling effects being produced by varying the diameter and length of the nozzle throat. The smallest degree of throttling, and hence the smallest plenum pressure rise, occurred when the contact diameter was less than that of the nozzle throat, whilst the highest degree of throttling corresponded to the longest length of parallel throat section which led to delayed relief from contact throttling. Their results highlight the danger of over-throttling in an attempt to increase pressurisation, since although the sought-for improvement in on-load pressure may be achieved, too severe a decompression has to be tolerated during the arc-recovery period. The resulting pressure transient is similar to that shown in Fig. 4.1a for a 2-pressure interrupter. Nonetheless, several workers (e.g. Tominaga et al., 1979; Yoshioka et al., 1979) have successfully achieved increased pressurisation without the penalties of severe decompression, possibly because they have operated in the EHV-type regime with powerful piston drive and large piston volume.

Concerning the shape of the piston chamber, Yoshioka et al. (1979) claim that, for a fixed piston-chamber volume and piston drive power, a short stroke rather than a small chamber cross-section produces a higher and an earlier pressure rise. Ueda et al. (1982) have considered the effect of separating the compression chamber into separate arc and piston chambers. Their results imply that the cross-section of the throttling interconnection between the arc and piston chambers may be optimised so that a significantly reduced piston-chamber volume, together with a low compression ratio and hence piston drive power, can produce acceptable pressure elevations during the arc-recovery period.

The pressure transients in a self-pressurising interrupter with suction (Fig. 3.2c) have been studied by Murai et al. (1981). The pressure transients in such interrupters are similar to those observed by Yoshioka and Nakagawa (1978, 1980) whereby the recovery-period pressure may be less than that during the peak current phase. A typical result, reproduced in Fig. 4.1e, shows the slow growth in arc-induced pressure which is produced immediately after contact separation. Substantial pressure growth is delayed until the arc length extends to the nozzle throat and the electrical power dissipated in the arc chamber is a maximum. As in the true puffer case (Fig. 4.1b), the magnitude of the pressure transient increases strongly with peak current. Also shown in Fig. 4.1e is the pressure reduction in the suction chamber on load and under no load. Although the pressure reduction is less on load, suction is clearly maintained during the arc-recovery period even for a peak current of 40 kA. Murai's results show the importance of having a sufficiently large initial volume for the suction chamber in order to prevent undesirable over-compression at the higher peak currents. Provided this condition is satisfied, mass flow rates from the arc chamber can be increased by an order of magnitude at low currents (~ 100 A RMS) without prejudice to the mass flow rate at higher current levels (~ 40 kA).

The possible use of SF_6-N_2 mixtures in puffer interrupters requires an appreciation of how the compressive action of piston and arc may each be

Characteristics of SF_6 interrupters   29

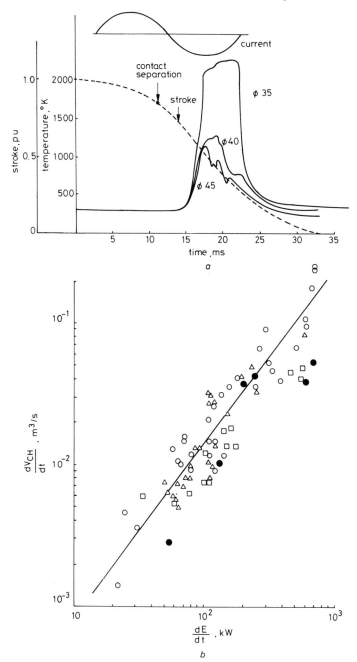

Fig. 4.2 *Arc-induced gas heating*
a Gas temperature at nozzle inlet for various diameters (Fig. 2, Ikeda et al., 1984. Copyright IEEE)
b Thermal bubble volume as a function of arc power input (Fig. 7, Sasao et al., 1981. Copyright IEEE)

30 Characteristics of SF_6 interrupters

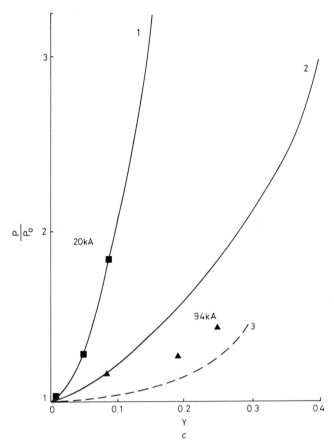

Fig. 4.2 *(continued)*
c Pressure elevation as a function of parameter Y for different breakers (Fig. 4c, Jones, 1984)
1 complete gas mixing 2 No mixing 3 SF_6 puffer
■ SF_6 rotary arc ▲ SF_6 two pressurre interrupter

modified. The adiabatic compression and decompression of an SF_6–N_2 mixture increases with nitrogen content owing to the dominant influence of the specific heat ratio (1·4 for N_2, 1 for SF_6) (Sasao *et al.*, 1982). Thus, for a perfectly sealed piston chamber, the pressure rise increases with nitrogen concentration, whereas for a loose piston with some gas leakage the pressure rise decreases with nitrogen concentration. However, at elevated temperatures, the pressure rise is insensitive to nitrogen concentration since the specific-heat ratios for N_2 and SF_6 are similar. The pressure transients in a puffer interrupter for various nitrogen concentrations and for on- and no-load conditions (Fig. 4.1*f*) show the effect of these compressive influences. Accordingly, the no-load compression decreases with an increase in nitrogen concentration whereas the on-load compression increases. Too high an N_2 concentration leads to a catastrophic decompression by the recovery period, which convincingly illustrates why the puffer principle was commercially inadequate using air as indicated above. The introduction of a hot-gas vent through the fixed contact passing through the interrupter nozzle makes the pressure transient less sensitive to nitrogen concentration at the expense of reduced pressure elevations. These results highlight the fact that puffer interrupters using SF_6-N_2 mixtures are even more critically sensitive to the interrupter geometry than SF_6 puffers, even on the basis of compression considerations alone. It is therefore not surprising that, as indicated by Sasao *et al.* (1982), there are conflicting reports in the scientific literature concerning the suitability of such mixtures for circuit-interruption applications.

Pressure transients produced in rotary-arc interrupters have been less well reported in the scientific literature. Nonetheless significant pressure changes may be produced depending upon the interrupter geometry, and significant aerodynamic resonances are also observable (Spencer, private communication). However, unless the rotary arc is being employed in a hybrid interrupter (e.g. Fig. 3.4*a*), where convection through a nozzle is required to dominate, there may be good commercial reasons for limiting such pressure elevations if the interrupter action relies dominantly upon cross-flow convection.

4.2 Thermal characteristics

The generation of an adequate pressure elevation is a necessary but not sufficient prerequisite for proper puffer action. In addition, controlled heating of the interrupter gas may be utilised to achieve the required compression, and advantage taken of any improved arc-control and quenching properties as outlined in Chapter 2. However, overheating to achieve the required compression at the expense of a serious degradation in the insulating properties due to excessive nozzle throttling by the contact, or excessive nozzle ablation, must be avoided.

4.2.1 Gas heating
Although the problems associated with gas heating are obviously well recognised, only indirect estimates are available of the extent of gas heating which occurs under conditions of practical interest, mainly on account of the experimental difficulties in making direct measurements. Below a critical value of nozzle-throat diameter, a dramatic increase in the gas temperature occurs up to 2000 K which may well assist gas compression, but whose persistence during the arc-recovery period is undesirable for EHV applications.

For EHV interrupters, lower temperature elevations are probably required during the recovery phase than in distribution interrupters in order to meet the demands of higher restrike voltages.

Ikeda *et al.* (1984) have investigated the gas heating in the converging upstream volume of a partial duoflow 300 KV interrupter. Fig. 4.2*a* shows their results for the influence of nozzle-throat diameter and point-on-wave contact separation. Point-on-wave contact-separation results show that the persisting temperature elevation during the recovery period is greatest in an intermediary time domain, when arcing at a peak current (with increased power input) has occurred whilst simultaneously the nozzle throttling effect of the contact has not been relieved. Below the minimum time boundary of this domain, the arc length at peak current is small enough to prevent overheating, whilst above the maximum time boundary, there is sufficient time for the nozzle throttling by the contact to be overcome to release the trapped thermal energy.

The approach of physically separating the arc and piston chambers into separate entities (Ueda et al., 1982) may be utilised to provide better control of arc heating, at least for distribution-voltage puffers. Typically an optimum value of 30–50% for the ratio of the cross-section of the separating constriction to the piston-chamber cross-section is claimed to limit temperature rise in the arc chamber to a broad minimum, and a relatively low value of piston-chamber cross-section (value dependent on operating conditions but not specified by the authors) can be tolerated without greatly affecting the temperature rise. Compression ratios (i.e. ratio of closed to open piston-chamber volumes) up to approximately 3 are claimed to produce no significant increase in arc-chamber gas temperature, even if the piston chamber volume is halved. Taken in conjunction with the pressure predictions for such a system (Section 4.1), Ueda et al. (1982) claim that optimum conditions can be achieved with a reduced-volume puffer cylinder and low piston drive power.

In the case of self-pressurisation with suction (Murai et al., 1981), hot gas is sucked into the suction volume during operation. The resulting temperature rise increases inversely with the suction-volume size, and so sets a lower limit to its value. For puffer interrupters utilising SF_6–N_2 mixtures (Sasao et al., 1982) the temperature rise in the piston chamber increases with nitrogen content so that performance attainment under extreme fault conditions would be more difficult to achieve than with pure SF_6.

The estimates of gas heating made by the above authors requires assumptions to be made about the degree of mixing which occurs between hot and cold gases. Unfortunately, it is difficult to ascertain from the information given by the authors what assumptions have been made. These will obviously have a significant effect upon both temperature and pressure predictions.

Results obtained by Shimmin (private communication) for a 420 kV puffer interrupter provide some clarification. A comparison has been made with the measured on load pressure transient of predictions based purely on the blocking effect alone of the arc (without any additional heating) and predictions assuming the formation of a heated volume which does not mix with the cool gas. In the latter case, compression is produced by the expansion of the heated-gas volume. The size of the thermal volume may be related to the rate of electrical-energy input using the empirical results of Sasao et al. (1982) (Fig. 4.2b). For a peak current of 36 kA, the measured on-load pressure transient compares favourably with the heating but no-mixing case, whilst the 'no heating' calculations predict excessive pressures during the stroke after completion of arcing.

This comparison confirms the occurrence of a more rapid exhaust of gas when heated, and shows that the accompanying decompression tendency observed with the 'distribution'-type puffers is also evident with EHV puffers. However, in the latter case, the decompression is insufficient to produce a pressure reduction because of a longer stroke, a more powerful piston drive and a larger piston volume.

It is, of course, possible that situations may exist when complete rather than no mixing of the heated and cool gases occurs. This could, for instance, occur when there is excessive energy input in a limited volume. Jones (1984) gives a comparison of the pressure elevations produced by both cases, on the assumption of no gas leakage, with typical changes measured in various types of interrupters. The induced pressure changes may be related to a simple scaling parameter

$$Y = (I_0 j_0 (1 - \alpha) l_0 / (\sigma \omega U a_3)$$ (4.1)

where I_0 is the peak current, j_0 the arc-current density at 1 bar absolute ($\sim 10^7 \text{Am}^{-2}$ for SF_6), $(1 - \alpha)$ is the fraction of radiated power absorbed by the gas (65–90% depending upon metal vapour contaminated), l_0 is the arc length contributing to the heating, σ the electrical conductivity, ω the angular frequency of the current waveform, U the volume of the pressurised part of the interrupter and a_3 is a numerical constant. This scaling parameter therefore embodies the influence of the interrupter geometry (l_0, U), the arc channel [j_0, $(1 - \alpha)$, σ] and the current waveform (I_0, ω), and so provides an approximate but convenient method for assessing the influence which particular changes in any of these parameters would have upon the pressure rise. Results for pressure elevation at the end of a single half-cycle of arcing as a function of Y show that (Fig. 4.2c), for $Y < 0.4$, complete mixing of heated and cool gas produces a greater pressure rise than no mixing and, of course, with a lower average gas temperature. A comparison of these predictions with some measured pressure changes in different types of interrupters (Fig. 4.2c) suggests that at limited current levels EHV puffers utilise no heated gas with no mixing, but with significant leakage through the nozzle. Interrupters utilising the rotary-arc principle utilise thorough mixing of the heated gas, which is more efficient in producing a greater pressure rise for a given value of the scaling parameter Y. The use of arc rotation in hybrid-type interrupters to improve pressurisation would therefore appear to be well founded.

4.2.2 Nozzle ablation

Puffer interrupters are more prone to severe nozzle ablation than 2-pressure interrupters on account of their reliance upon a higher degree of nozzle blocking to provide compression. In some types of interrupters deliberate use is made of ablation from an insulating nozzle, but it is unlikely that excessive ablation of a metallic nozzle would assist interrupter performance. However, a knowledge of the ablation from metallic as well as insulating nozzles is relevant to interrupter design, since, although increasing use is made of insulating nozzles (almost entirely PTFE at present), nonetheless hollow metallic contacts are used as subsidiary nozzles in partial duoflow interrupters (Fig. 3.1c). For these reasons the improved understanding of ablation which is emerging from fundamental studies is of value to circuit interruption (e.g. Niemeyer, 1978; Jones and Fang, 1980; Kovitya et al., 1978; Noeske, 1977; Noeske et al., 1983; Kirchesch and Niemeyer, 1985; Jones et al., 1986).

Energy for inducing ablation is available by thermal conduction if the arc plasma makes contact with the nozzle wall. However, this is not a prerequisite for ablation since sufficient radiated power is often available from the arc even when it is remote from the nozzle wall. For a nozzle fully exposed to a whole half-cycle of sufficiently high-current arcs, there is strong evidence that radiative transfer is responsible for initiating nozzle ablation (e.g. Jones and Fang, 1980).

For metallic nozzles, the thermal diffusivity is sufficiently small for heat penetration in the solid nozzle to be limited to a depth of only 0·5 mm during a half cycle at 50 Hz (Fang and Newland, 1984). Consequently most of the absorbed heat is available for ablation. For PTFE nozzles (which are partly transparent to intense radiation) optical measurements show a non-uniform absorption with depth, so that the absorption of most radiation is highly localised close to the surface and promotes ablation more readily than would otherwise be expected. However, ablation of PTFE is difficult to define since it is governed by the formation of chemical fragments due to the fracture of various molecular bonds, the particular bonds involved depending upon the spectral content of the radiation (Newland, private communication). Nonetheless, measurements indicate that for a given nozzle geometry PTFE ablates at a lower peak current than a metallic nozzle.

Not surprisingly, the most severe ablation occurs in the vicinity of the nozzle throat where the radiative flux density at the wall is most severe owing to the closer proximity of the wall to the arc. Also, as expected, the amount of ablation increases with fault current and with multiple half-cycle arcing. The increase of ablation with peak current is considerably more pronounced for a full cycle of arcing that for a single half-cycle.

The onset of ablation can occur early during the current half-cycle even for peak currents well below the maximum rated value for the nozzle (Fig. 4.3a) particularly if the arc is contaminated with more highly radiating contact vapour. Combined spectroscopic and local pressure measurements (Jones et al., 1986) indicate that it is only a small fraction of the ablated mass which is locally resident in the nozzle at any instant (Fig. 4.3a), the larger fraction having been convectively transported through the nozzle. Throat-pressure measurements (e.g. Taylor, 1983; Noeske et al., 1983) can indicate the instant beyond which a substantial part of this mass is transported back into the piston chamber to produce additional pressure changes in the piston chamber. It is not only the generation of ablated mass which produces this backflow but also additional heating of the resident ablation products due to further absorption of radiation from the arc. In the case of a metallic nozzle, spectral-absorption measurements show that this heating of peripheral ablated mass may persist well after peak current, so that vapour at temperatures of about 2000 K would not be unreasonable during the recovery period (Fig. 4.3b).

A comparison of the ablation in a monoflow and a duoflow interrupter

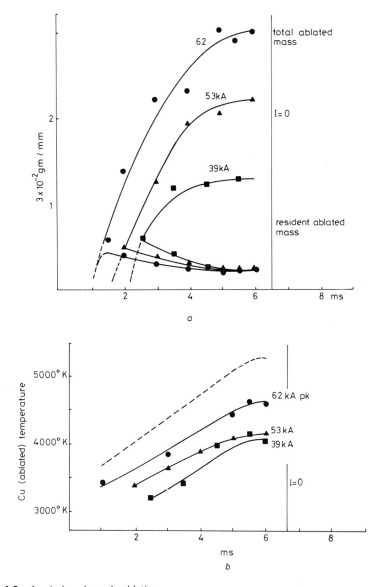

Fig. 4.3 Arc-induced nozzle ablation
  a Total and resident ablated mass (Shammas and Jones, private communication)
  b Temperature of ablated material (Shammas and Jones, private communication)

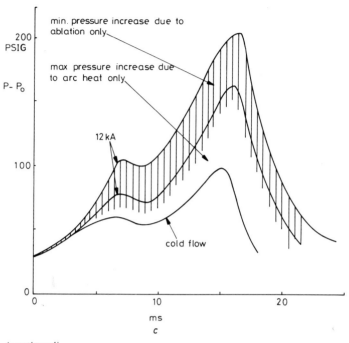

Fig. 4.3 *(continued)*
c Contribution of ablation to the pressure elevation (Fig. 14, Noeske, 1977. Copyright IEEE)

employing insulating nozzles of identical geometry shows that the throat ablation is identical for both nozzles, whilst the downstream ablation is twice as big for the monoflow nozzle (Noeske *et al.*, 1983).

Under real puffer conditions, the development of nozzle ablation is more complicated than a simple exposure to the full half-cycle current arc since the main PTFE nozzle throat and downstream sections are exposed to the arc for less time than the upstream nozzle section on account of the protection afforded by the moving contact (e.g. Ali *et al.*, 1985; Noeske *et al.*, 1983). Taking account of such effects, Noeske *et al.* (1983) estimate that, for an assumed temperature of 2000 K and a CF_4 composition for the ablation products the ablated mass if transported back into the piston chamber would occupy about one-third to one-half of the chamber volume with a corresponding effect upon the pressure rise. A comparison of the estimated contributions made to the pressure elevation by gas heating and ablated mass for advanced nozzle blocking (Fig. 4.3c) shows that the ablation contribution cannot be neglected.

Commercially, the possibility of scaling interrupter designs is obviously attractive, although the evolution of scientifically rigorous scaling laws is a formidable task on account of the conflicting effects of the many processes involved. Nonetheless, progress has been made with scaling laws for geometries (e.g. Fang *et al.*, 1980), and considerations made in Section 4.2.1 indicate approximate procedures which could take account of gas-heating effects. Noeske et al. (1983) have extended such scaling considerations to take account also of nozzle ablation. Thus, in order to produce similar pressure transients on and off load in puffers of different geometrical size, successful account has been taken of the influence of nozzle ablation in scaling the acceptable fault current. It is assumed that the thickness of the ablation layer at the nozzle wall is independent of the nozzle size as long as the effective current density is fixed. With an ablation-layer thickness of 1 mm it is shown that, for the particular nozzles of interest, where the linear dimensions of the nozzles are halved the current would need to be reduced by 12% more than if ablation were neglected; whilst for a reduction of a third in the linear nozzle dimensions the current would need to be 23% less than the no-ablation value. The approximate validity of the proposed scaling is supported by pressure transient measurements which show encouraging agreement for currents scaled in the range 17–72 kA RMS. Clearly, any comparison of performance between scaled puffer interrupters needs to take full account of such ablation effects.

4.3 Electromagnetic characteristics

For convenience, the electromagnetic characteristics of circuit-breaker arcs may be divided into two categories corresponding to those associated with the internal properties of the arc plasma and those concerned with the overall behaviour of the arc column as an entity. Internal electromagnetic phenomena are manifest as plasma compression or circulating plasma flows. At sufficiently high currents with axisymmetric arcs, the simple plasma compression and linear acceleration (Fig. 4.4a) observed by Maecker (1959) is complicated by a swirling plasma (Fig. 4.4a) (Jones et al., 1982; Shishkin and Jones, 1985) with rotational frequencies of several tens of kilohertz. Additionally, vortex flows around the periphery of the arc column close to the upstream contact may be generated owing to the contraction of the arc column at the arc root (Fang and Newland, 1984). For axisymetric arc columns in crossflow, complex circulating plasma flows may be generated (e.g. Jones and Fang, 1980).

Despite the occurrence of such complex plasma flows during the high-current arcing phase, there appears to be no indication of any deleterious arc-control or quenching effects, the latter because the time scales of the flows are too short compared with the current-decay rates involved in circuit interruption.

Unlike the internal electromagnetic effects, the external effects have important implications for circuit interruption. For instance, in the case of a duoflow interrupter with metallic nozzles which also act as contacts (Fig. 3.1c), the arc is subjected to a magnetic force whose direction and magnitude are governed by the geometry of the current-carrying metallic nozzles. Small diameters of the remote nozzle sections produce an outward force which would oppose the movement of the arc into the nozzles by the inward flow (Fig. 4.4b). However,

38 *Characteristics of SF_6 interrupters*

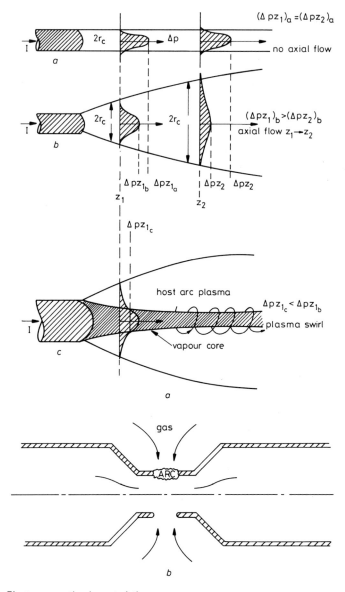

Fig. 4.4 *Electromagnetic characteristics*
a Self-magnetic effects (Fig. 1, Jones et al., 1982)
b Magnetic forces in duoflow geometry (Fig. 1, Hurley, 1973. Copyright IEEE)

for sufficiently large nozzle diameters the magnetic force may reverse and assist the flow in transferring the arc into the internozzle gap (Hurley, 1973).

The most obvious manifestation of external electromagnetic influences is in the rotary-arc interrupters, which rely upon the production of a Lorentz force

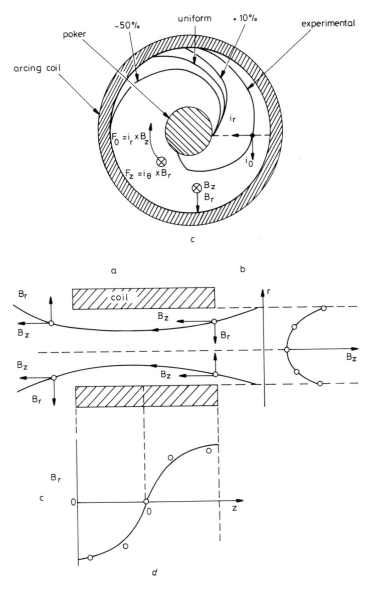

Fig. 4.4 *(continued)*
c Rotary-arc shape (Fig. 5, Spencer et al., 1985)
d Spatial variation of magnetic flux density (Fig. 5, Spencer et al., 1985)

from the cross-product of the magnetic flux density and fault current to rotate the arc. In the Hawker Siddeley type of interrupter using a rod contact and a peripheral contact which also supports the coil (Fig. 3.3a), the arc is exposed to a mainly axial magnetic field. In such a configuration the arc assumes a curved,

40 Characteristics of SF_6 interrupters

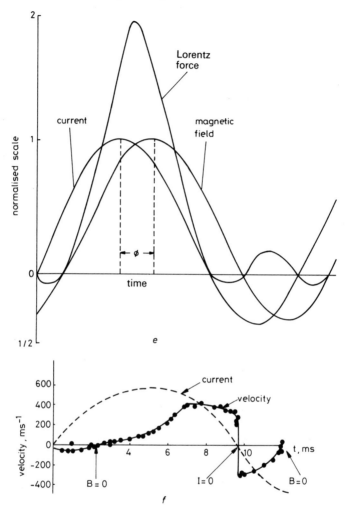

Fig. 4.4 *(continued)*
e Resultant Lorentz force (Spencer, private communication)
f Rotational velocity of arc (Spencer, private communication)

rather than simple radial, shape (e.g. Spencer *et al.*, 1985; and Fig. 4.4c), the precise shape being determined by nonuniformities of the magnetic-flux distribution and arc-root influences. As a result of this convoluted shape and the axial and radial components of the magnetic flux (e.g. Fig. 4.4d), the arc is not only rotated azimuthally (owing to $F_\theta = j_r \times B_z$) but also displaced axially (owing to $F_z = j_\theta \times B_r$). Thus the arc penetration is limited by the fact that the radial component of the flux density reverses with axial position through the coil (Fig. 4.4d). The radial expansion of the helical arc column within the coil volume leads to the arc short-circuiting its axial penetration, so producing fluctuations in the overall arc length (Spencer *et al.*, 1985) during a current half-cycle.

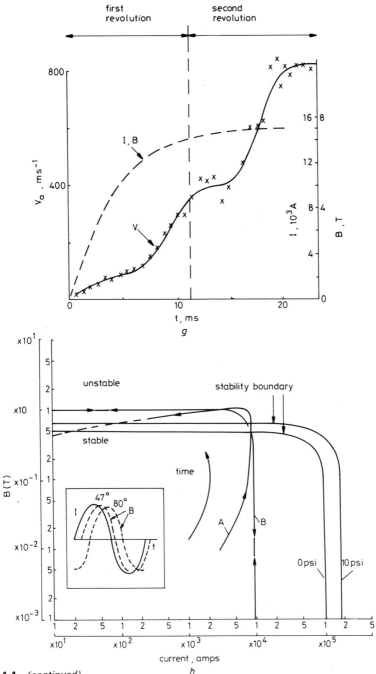

Fig. 4.4 *(continued)*
g Effect of gas heating on arc velocity (Kopainsky and Schade, 1979)
h Electromagnetic stability (Fig. 6, Spencer et al., 1985)
A − 47°; B − 80°

42 Characteristics of SF₆ interrupters

The annular contact of this type of interrupter, which also acts as the coil yoke, carries eddy currents which have two important effects. The first is to act as a shield which reduces the magnitude of the magnetic flux density. The extent of this reduction depends upon the material of the contact and increases with its thickness (Turner and Chen, 1985). Thus a compromise is required between an acceptable thickness to withstand arcing and provide sufficient mechanical strength to withstand the magnetic compressive forces, and a thin yoke to avoid too much reduction in magnetic flux density.

The second effect of the eddy currents in the coil yoke is to produce a phase lag between the fault current through the arc and the magnetic flux density. There is therefore a delay in the growth of the Lorentz force (Fig. 4.4e) and in the rotational acceleration of the arc column. Clearly, this phase lag between fault current and magnetic flux density needs to be optimised to provide the best arc control for current interruption. The resulting arc motion is complicated by the fact that the Lorentz force may reverse for a short period after current zero if arc reignition occurs (Fig. 4.4f), which, because of strong drag forces, causes a temporary reversal of arc movement. On the other hand, if the arc rotation is sufficiently rapid, the arc column may move in its own heated wake after the first revolution and be accelerated owing to a reduction in the drag consequent upon the lower density of the heated wake (Fig. 4.4g; Kopainsky and Schade, 1979). The electromagnetically induced arc dynamics are further complicated by the point-on-wave contact separation (Turner and Chen, 1985) and by the degree of assymetry of the current waveform. For instance, with two-thirds current assymetry, the rotational velocity of the arc column during the minor loop may be substantially reduced on account of the direction of the driving Lorentz force being reversed in relatively rapid sequence without the opportunity to grow to a magnitude comparable to the symmetric-current-waveform case.

The existence of electromagnetic forces can also influence the stability of the arc column (e.g. Ragaller, 1974; Schrade, 1973). Fig. 4.4h shows the stability boundary separating stable and unstable arc operation in terms of the relationship between magnetic flux density and arc current according to the relationship (Schrade, 1973):

$$8\pi p_\infty - B_0^2 - 8\mu_0^2 I^2/R^2 \begin{array}{l} > \text{ stable} \\ = 0 \text{ critical} \\ < \text{ unstable} \end{array} \quad (4.2)$$

where R = arc radius
B_0 = magnetic flux density
p_∞ = gas pressure
I = arc current

The growth of instabilities is most likely along the direction of the imposed magnetic field, i.e. perpendicular to the direction of rotation and axially into the coil in the case of the Hawker Siddeley rotary-arc interrupter. The vertical branch of the stability boundary (Fig. 4.4h) is governed by the dominance of self-magnetic effects whilst the horizontal boundary is determined by applied magnetic fields. Thus, although currents in excess of 200 kA would be required for instability due to self-magnetic effects, an applied magnetic flux of only 700 mT would be sufficient to produce instability in arcs carrying a current of only a few amperes.

As an example, also shown in Fig. 4.4h are the characteristics for a 9 kA peak current and a magnetic flux density of 113 mT/kA but with different phase lags (47° and 80°). Whereas in the case of the 47° lag the arc only becomes unstable after current peak and becomes stable again before current zero, in the case of the 80° phase lag the arc remains unstable throughout the entire current loop apart from a short period close to current peak. Nonetheless, the importance of the stability condition of the arc for circuit-interruption applications still remains to be resolved.

Although much progress has been made in understanding some of the electromagnetic characteristics of SF_6 interrupter arcs, there are several aspects which require further investigation and which could well lead to developments that could be of importance for new circuit-interruption applications.

4.4 Thermal-recovery characteristics

The thermal-recovery period of SF_6 interrupters persists for a short time of the order of 4 μs either side of the current zero at the end of the current wave. Because of the short time duration involved, it is predominantly the condition and behaviour of the contracted arc column (Diameter $\sim$ 1 mm), rather than processes remote from the arc, which dominantly govern the thermal recovery. Consequently, the thermal recovery in the different kinds of interrupters is expected to be governed by similar considerations and processes. A possible exception may be the distinction between axisymmetric (2-pressure and puffer type) and asymmetric-(rotary) arc interrupters. The relevance of the type of interrupter is in dictating the manner in which this required common thermal-recovery condition of the arc column is evolved from the higher current period of arc control. Detailed research into the thermal-recovery period has therefore often been undertaken under the most ideal operating conditions which are compatible with circuit-interrupter operation.

4.4.1 Axisymmetric (puffer-type) interrupters
There have been several investigations of the dependence of thermal recovery of gas-blast arcs upon nozzle properties (size, geometry, material, single or double) contact properties (size and position), and gas (pure SF_6 and mixtures, various pressures) (e.g. Noeske *et al.*, 1983; Noeske, 1977, 1981; Wang *et al.*, 1982; Frind, 1978; Lewis *et al.*, 1985, Briggs and King, 1979). Such investigations

avoid complications associated with contact movement and often utilise synthetic current waveforms (e.g. ramps or high-frequency sinusoids).

4.4.1.1 Nozzle and contact considerations

For monoflow conditions the upstream contact material can have an effect upon thermal recovery at high currents when significant contact evaporation occurs. Graphite gives the best performance, followed by copper–tungsten (e.g. Frind, 1978). Thermal recovery is severely retarded if this upstream contact is positioned closer than a distance of approximately the nozzle radius from the nozzle throat (Noeske et al., 1983). The presence of fingers in an otherwise unchanged contact shape has no major effect upon the recovery provided the nozzle is correctly dimensioned (Noeske et al., 1983).

Nozzle-geometry optimisation is concerned not only with the production of an aerodynamically adequate flow but also with generating a flow which accelerates the thermal recovery of the arc. Thus the flow cross-section at large distances from the nozzle axis is not important (for thermal-recovery speed); sonic flow injection in a 2-nozzle system is detrimental to recovery speed; flow stabilisation using a constant-diameter throat section benefits recovery speed; and the formation of a substantial stagnant region upstream (due to the presence of contacts, or nozzle geometry problems) reduces recovery speed. It may also be assumed as the simplest possibility that the subsonic and supersonic sections of the nozzle, separated by the Mach 1 region at throat (which may be a complicated 2-dimensional surface), can be treated and optimised separately.

For monoflow conditions Noeske et al. (1983), recommend an inlet angle of 45° followed by a parallel throat of limited length (Fig. 4.5a). For the downstream nozzle section, small expansion angles of less than 15° are not recommended because of their tendency to prolong the throttling effect of the contact moving relative to the nozzle. A bell-shaped downstream nozzle section has been proposed (Fig. 4.5a), which is the optimum shape for accelerating the recombination of ions and electrons into neutral atoms in the flow of ionised gas following arcing (Noeske et al., 1983). The thermal recovery associated with this new nozzle has been shown to be about 30% better than with the more conventional uniform-divergence sections. Moreover, the new nozzle shape provides faster deblocking and directs the exhaust flow of hot gases. The throat diameter of the nozzle is determined by the maximum fault current. In the case of 2-pressure interrupters, the threshold for nozzle blocking could be used as a design criterion (e.g. Fang et al., 1978). For puffer interrupters which rely upon nozzle blocking for their operation, current densities up to $12.5 \, kA/cm^2$ have been tolerated, and still provide acceptable thermal recovery (Noeske et al., 1983).

Nozzle-geometry scaling under such conditions needs to take account of the layer of ablation material at the nozzle periphery, as already discussed in Section 4.2.2. Furthermore, the nature of the nozzle material has an influence under such conditions. Comparison tests with different nozzle materials (Wang et al.,

1982) suggest that there is little difference between copper, steel and carbon nozzles, but that PTFE leads to a somewhat inferior performance. The reason for this poorer behaviour of PTFE compared with copper and steel is not so much associated with the influence of the ablation products but rather with changes which occur in the shape of the nozzle throat following the removal of material by ablation. In the case of copper and steel, there is significant erosion of the throat so that performance at a given high current improves with early aging. With PTFE, no such overall change occurs on the same time scale, but rather an increase in the pitting of the nozzle surface. Tests before the onset of such effects reveal little performance difference between PTFE and copper nozzles. More recently (Saito and Honda, 1981) evidence is claimed for improved performance when PTFE is replaced by a type of glass-filled PTFE.

In the case of duoflow conditions, the nozzle-throat and downstream-nozzle divergence criteria are identical to the monoflow case. For the upstream inlet region, however, an inlet-wall radius of about 37% of the throat diameter is recommended (Noeske et al., 1983), with a gap between the two nozzles of 80% of the throat diameter.

The performance improvement attainable with optimised duoflow over an optimised monoflow nozzle may be described in terms of an 'improvement

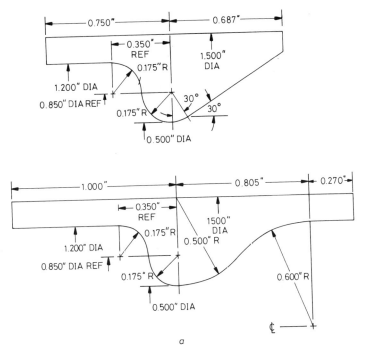

Fig. 4.5   *Thermal-recovery characteristics of puffer interrupters*
     a *Monoflow nozzle profiles (Figs A-21, 22, Noeske et al., 1983)*

46 Characteristics of SF_6 interrupters

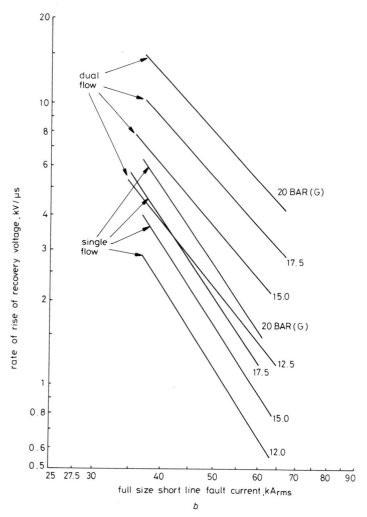

Fig. 4.5 *(continued)*
b Projected performance of duoflow nozzles (Fig. H-19, Noeske et al., 1983)

factor', $2(1 + \alpha)\beta$ (Noeske *et al.*, 1983). The parameter α is a measure of the performance improvement of a monoflow nozzle which could be achieved through the elimination of contact vapour. The parameter β takes account of the different geometrical parameters of the mono- and duoflow nozzles, which are inevitable upstream and which may affect not only the local contribution to interruption, but also the flow, and hence influence upon interruption of the downstream-nozzle section.

β may be greater or less than unity depending upon whether or not the duoflow nozzle can be optimised to a degree which gives it an interruption

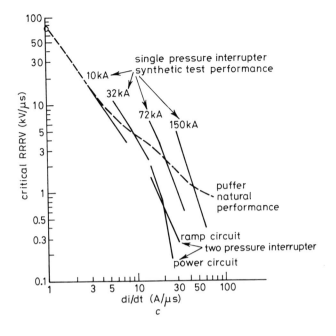

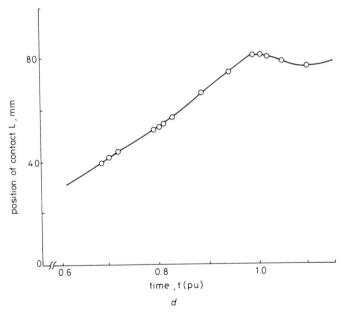

Fig. 4.5 *(continued)*
c Comparison of 2-pressure and puffer performance (Fig. 6, Shimmin et al., 1985)
d Successful interruption variation with contact separation (Fig. 7a, Yoshioka and Nakagawa, 1978. Copyright IEEE)

48 Characteristics of SF_6 interrupters

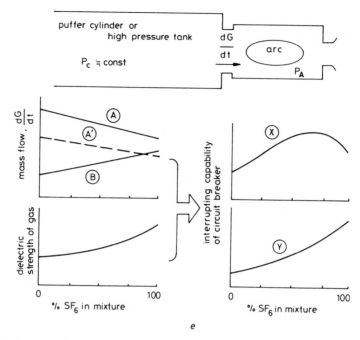

Fig. 4.5 *(continued)*
e Performance of SF_6 mixtures in open and closed geometries (Fig. 9, Sasao et al., 1982. Copyright IEEE)

performance either superior or inferior to that of two monoflow nozzles without electrode vapour. The factor 2 takes account of the fact that the duoflow nozzle consists of two units each with an improvement factor of $(1 + \alpha)\beta$.

For equally optimised mono- and duoflow nozzles, the improvement factor should tend asymptotically to 2 at low currents when contact-vapour effects should become less important in the monoflow case. Improvement factors of 1·9 have been quoted for such conditions (e.g. Noeske *et al.*, 1983). For higher currents with large puffer pressures, the improvement factor increases to 3 (Noeske *et al.*, 1983; King, 1978) because the contact-vapour effects become more pronounced, and these disproportionately reduce the monoflow performance. Scaling with such an improvement factor leads to the projected performance characteristics shown in Fig. 4.5*b*. In the case of a partial-duoflow interrupter, a scaling factor of 1·4 appears to be justified (Noeske *et al.*, 1983).

The above discussion provides general guidelines concerning idealised nozzle-geometry effects. The simpler aspects of the geometry may be incorporated into an empirical description as given by Ancilewski *et al.* (1984) and which allows departures from ideal behaviour to be identified, which, for economic reasons, may need to be commercially tolerated. This empirical relationship relates the rate of rise of recovery voltage to the current-delay rate di/dt, gas pressure p, current-zero arc area A, Mach number M, throat diameter D_t and rate of rise

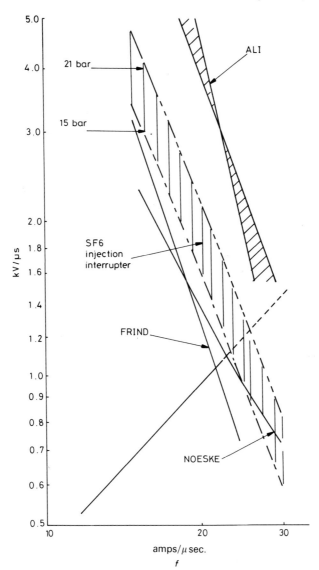

Fig. 4.5 *(continued)*
f Predicted performance of liquid SF_6 injector (Fig. 6, Briggs, 1985)

of recovery voltage (RRRV)

$$RRRV = aLd_tMpq(di/dt)^{-1}A^{-1} \qquad (4.3)$$

The current-zero arc area is explicitly introduced because Ancilewski *et al.* (1984) argue that, if the radial components of flow velocity are less than the rate of collapse of the arc cross-section, then the greater arc area A at current zero leads to a more rapid reduction of RRRV with di/dt for a given geometry (L, D_t fixed). The improved performance obtained with sonic rather than subsonic flow is taken into account via the Mach number M. The linear dependence of RRRV upon the ratio of throat diameter to upstream gap is consistent with the results given by Yanabu *et al.* (1981), but appears to be limited in range of application if the results of Noeske *et al.* (1983) are taken into account.

Shimmin *et al.* (1985) have extended the above type of empirical description to cover more extensive operating conditions including those when self pressurisation may dominate over piston compression:

$$\text{RRRV} = ap^{1\cdot5}(1 + 1\cdot7\Delta p)A_0^{-1\cdot5}(1 + 5 \times 10^{-5}I(di/dt)^2)^{-1} \quad (4.4)$$

$$\underset{\substack{\text{piston}\\\text{pressure}}}{} \underset{\substack{\text{arc}\\\text{induced}\\\text{pressure}}}{} \underset{\substack{\text{low current}\\\text{factor}}}{} \underset{\substack{\text{peak current}\\\text{factor}}}{}$$

Geometry effects are absorbed into the parameter a. This relationship indicates the well established fact that there is a basic decrease in performance through the dependence upon the current zero arc cross-section A_0, which can be shown to be identical with the predictions of the Mayr theory. It has also been well established that thermal recovery can be improved if the background pressure is increased. Many researchers have found the dependence to be with regard to $p^{1\cdot5}$ (e.g. Frind and Rich, 1974; Garzon, 1976; Airey *et al.*, 1976). However, in the puffer interrupter, the current-zero pressure is the value which dominates performance (e.g. Shimmin *et al.*, 1985), and this effect should rigorously be taken into account separately from the background pressure since it depends, via the peak current, upon di/dt (Shimmin *et al.*, 1985). At very high peak currents with 2-pressure interrupters, there is a more rapid degradation of performance (Fig. 4.5c), which Lewis *et al.* (1985) have identified to be due to a disproportionate change in the plasma temperature of the arc-column compared with arc-diameter changes. This influence is in addition to the effect of retarded radial inflow identified by Ancilewski *et al.* (1984). As a result of this influence, the dependence of puffer performance upon di/dt varies with peak current I such that at low I the performance increasingly becomes more severely dependent upon di/dt (e.g. Fig. 4.5c continuous lines) (Yoshioka and Nakagawa, 1978). The empirical relationship (Eqn. 4.4) also indicates that the degradation in performance which occurs as peak current is increased is limited in the puffer case (Fig. 4.5c) by the increased pressure at current zero due to the arc action. Furthermore, with properly optimised design these pressure increases may overcome the degrading effects of peak current, so producing a situation whereby performance can increase with peak current. This limit would, of course, correspond to the self-pressurising mode of operation, and neglects any dielectric-recovery effects.

4.4.1.2 Piston-travel and contact-throttling effects: There are also piston-travel and contact-throttling effects which need to be considered. Yoshioka and

Nakagawa (1978) have shown that the minimum time during the piston stroke at which interruption can be achieved is determined by the contact position within the main interrupter nozzle. This position is such that the effective flow cross-section around the contact is greater than the nozzle-throat area. This is also consistent with the conclusions of Yanabu et al. (1982). On the other hand, the maximum time at which interruption occurs is determined by the pressure in the puffer chamber. A typical illustration of the range of the stroke within which successful interruption can be achieved is shown on Fig. 4.5d. It is interesting to note that the pressure in the puffer chamber at the minimum arcing time was greater than that at the maxiumum arcing time. Yoshioka and Nakagawa (1980) have shown that, as a result of this interruption, the minimum time for interruption is longer for a nozzle with a longer parallel throat section and shorter for a nozzle with a clearance between the contact and the nozzle throat. Also a nozzle geometry with a large throat diameter was shown to have interrupting ability, despite a lower puffer pressure under on-load conditions. The interrupting ability declined for a current density of $5 \cdot 5$–$6\,\text{kA cm}^{-2}$ for the smaller type of interrupters used for distribution-system applications. Furthermore, the interrupting ability in a 1-cycle test was found to be higher than that in a half-cycle current test. In this case, increases in puffer pressure caused by arcing had a favourable effect upon interrupting ability. Some of these pressure effects appear to be different from the beneficial effects observed in high-voltage circuit breakers, possibly on account of the decompression effects discussed in Section 4.2.1 which occur in the smaller-volume distribution-system-type interrupters.

4.4.1.3 SF$_6$ mixtures: The possibility of using mixtures of SF$_6$ with other gases has intrigued researchers on account of the possibility of overcoming SF$_6$ liquefaction problems at low temperatures and high pressures. A comprehensive survey of the interruption performance of numerous SF$_6$ mixtures has been made by Noeske (1981), Lee and Frost (1980) and Kinsinger and Noeske (1976). The results of this survey illustrate the enormous range of performance which can be achieved. The observed trends have been explained in terms of a 3-zone arc model and an equation which describes the different energy carried by particles which diffuse across the boundary from the non-ionised but fully dissociated zone into the non-dissociated zone:

$$\Delta(\text{power loss}) = (C_d)\left[\alpha_x \sum_{n_x}(D_x N_x N_{Bx} E_{Bx}) + \alpha_y \sum_{n_y}(D_y N_y N_{By} E_{By})\right]\Delta t$$

(4.5)

where C_d is the sound velocity in the dissociated gas mixture; α is the mixture ratio for each gas x, y, where n_x, n_y is the number of atoms in each molecule, and x, y, is the diffusion coefficient of an atom of mass M_n against molecules of average mass M; N is the relative concentration of atoms with mass M_n; N_B is the number of bonds in the molecule; and $E_{Bx,y}$ is the average bond energy in each molecule X, Y. A degradation of recovery for C_2F_6 with about 20% SF_6 occurs and may be explained by the formation of strongly bonded diatomic species such as CS from the mixed gases. The formation and influence of such species is favoured by high pressure and high di/dt. The difference between some of Noeske's (1981) thermal recovery rates and those reported by Frind et al. (1977) and Lee and Frost (1980) is attributed to the formation of such species in the one case but not in the other owing to different operating conditions.

For practical applications, the different requirements for puffer interrupters (low sonic velocity with relatively low storage pressure) and for 2-pressure interrupters (gas storage at high pressure) need to be considered. SF_6 mixed with CF_4, C_2F_6, He and N_2 make interesting candidates for 2-pressure interrupters. Mixtures of He or C_2F_6 with low partial pressures of SF_6 have no liquefaction problems up to a pressure of 35·5 bar, and have similar recovery properties to SF_6 at 22 bar. Similar, although less pronounced, improvements may be achieved with CF_4 and N_2.

For puffer breakers, any mixture of SF_6 with C_2F_6 or CF_4 would be acceptable because of their similar mass. Because the storage pressure needs to be relatively low, higher ratios of SF_6 can be used compared with 2-pressure breakers. However, low SF_6 content in mixtures with He and N_2 is prohibited because of the resulting high sonic velocities, although Briggs and King (1979) have shown that performance improvements occur at high di/dt owing to a more rapid inflow component (Section 4.3.1.2). Application of the superior thermal recovery speed of H_2, D_2 and CH_4 is likely to be restricted by leakage, flammability and dielectric-breakdown problems.

The importance of the circuit-breaker configuration in determining performance with mixtures of SF_6 and N_2 has been considered by Sasao et al. (1982), and this could explain the discrepancies between the results of various investigators. Two types of arc chambers are identified, being called 'open' and 'closed' types. In the open type of chamber decompression is severe (Section 4.2.1), so that reinforced piston action is required. For the closed type of chamber, the compression increases with nitrogen content but at the expense of arc quenching due to the higher gas temperature. If separate arc and piston chambers are utilised (Fig. 4.5e), the decompression in the piston chamber is negligible. Thus, for an open-type arc chamber, the mass flow rate from the piston chamber decreases with SF_6 content, whilst the opposite is true for the closed-arc chamber (Fig. 4.5e). With a forced flow from the piston chamber the mass flow rate decreases as indicated by curve A because of a reduced pressure difference between the arc and piston chambers. Thus, for some cases of the open-type arc chamber, there will be better arc-quenching ability for a certain SF_6/N_2 ratio compared with pure SF_6 (X on Fig. 4.5e). This situation corresponds to the results reported by Grant et al. (1976), Garzon (1976) Lee and Frost (1980), Frind et al. (1977) and Noeske (1981). This condition for the highest arc-quenching ability corresponds to the case when the arc-induced pressure rise is dominated by the outward mass flow from the piston chamber, which is released at low current levels. On the other hand, with the 'closed-type' arc chamber, in which the pressure rise dominates the arc-quenching process, the highest arc-quenching ability occurs in the case of pure SF_6.

These considerations illustrate the additional complexities which need to be taken into account if mixtures with SF_6 are to be used.

4.4.1.4 Liquid SF_6 injection: Since thermal-reignition performance of an SF_6 interrupter improves as the operating pressure is increased, Garzon (1976, 1977) investigated the possibility of utilising pressures close to and above the liquefaction point of SF_6. The onset of the liquid phase of SF_6 was found to be accompanied by a dramatic improvement in performance.

More recently Briggs (1985) has explored the use of liquid SF_6 injection into the nozzle of a 2-pressure air-blast interrupter designed for 132 kV duty at 21 bar nominal air pressure. By injecting liquid SF_6 at a pressure of 56 bar to apply 7% SF_6 by volume, the $(dV/dt$ versus $di/dt)$ characteristic was measured. A comparison of the projected performance of the liquid SF_6 injector with that of conventional SF_6 interrupters reported by other researchers is shown in Fig. 4.5f. Briggs concludes that it is the mass of SF_6 injected which is the key parameter in governing the performance, and that its influence is primarily through local pressurisation rather than through the velocity of convection.

4.4.2 Assymetric (rotary arc) interrupters
Although the importance of various factors which govern the performance of a rotary-arc interrupter are still under investigation, detailed theoretical investigations (Spencer *et al.*, 1985) confirm that the phase angle between current and magnetic flux, the magnitude of the flux, and gas pressure all have a dominant effect.

The phase-lag angle between current and magnetic flux is given by (e.g. Zhang *et al.*, 1985):

$$\tan \theta = \frac{\omega L_T l_T \Delta}{\varrho_T \pi d} \qquad (4.6)$$

where ω is the angular frequency of the current waveform; L_T, ϱ_T are the self inductance and resistivity of the cylindrical contact material; and l_T, Δ, d are the length, thickness and diameter of the cylindrical contact. Empirically, optimum performance for the rotary-arc switch is found for $\theta \sim 45°$, which can be achieved by suitable choice of contact material and its thickness according to eqn. 4.6.

The effectiveness of the coil structure for providing sufficient magnetic flux is to cause current interruption governed by the parameter (Zhang *et al.*, 1985)

$$N = B_0/I_B < \frac{W \cos \theta}{0.786(L + 0.44D)} \qquad (4.7)$$

54 Characteristics of SF_6 interrupters

where B_0 is the magnetic flux density and I_B is the current to be interrupted. W, L, D are the number of turns, length and diameter of the coil, respectively. In practice, θ, and hence N, will vary within the confines of a given coil on account of non-uniformities at the coil ends.

Experiments indicate that the maximum current that can be interrupted increases fairly linearly with the nominal pressure within the interrupter chamber. Tests performed with a given interrupter yield the empirical relationship between recovery voltage (RRRV) and current decay rate (di/dt):

$$\text{RRRV} = ap^{1.5}(di/dt)^{-\alpha}B^{1.9}f(\theta) \tag{4.8}$$

This suggests that the background pressure (p) effect is similar for both puffer and rotary-arc interrupters. Also, whereas performance of a puffer breaker can be enhanced by arc-induced pressurisation, with a rotary-arc breaker an enhancement may be achieved by increasing the magnetic flux density. As arc-induced pressurisation increases with peak current for a puffer, so that magnetic flux density increases with peak current for a rotary-arc interrupter. As a result the performance deterioration as peak current is increased is less pronounced than at constant B. (Jones et al., 1985).

A similar, but not identical, arrangement to the rotary-arc-interrupter geometry has been proposed by Miyachi and Naganawa (1974) and Naganawa et al. (1985) for DC interruption. The arc is established between two contacts lying along the axis of two coils and forms outwardly expanding loops. The current-interrupting properties are governed by the radial-expansion speed of the loops and the ensuing total length of the arc column. The interrupting ability is assessed in terms of the overvoltage generated following interruption and the arc duration before interruption. Experiments show that, with SF_6, the arc duration decreases and the overvoltage increases as the current increases. However, the overvoltage may be limited at the expense of increased arc duration by using a mixture of SF_6 with nitrogen (Fig. 4.6). Tests with this type of interrupter have been limited to a maximum current of 350 A, so that its commercial feasibility at more realistic current levels remains to be explored.

4.5 Dielectric recovery

Following the thermal-recovery period, the SF_6 interrupter is required to recover dielectrically and to withstand voltages up to several hundred kilovolts within a few hundred microseconds of current interruption. This determines the voltage limit of a high-voltage breaker, which in turn governs the number of interrupter units to be connected in series in order to cope with the transient recovery voltage of a given network. Failure to recover dielectrically may be due to a number of factors, which include the retarded dielectric recovery of the remnant arc column, the trapping of overheated gas produced during the high-current phase and excessive local electric field strengths produced by geometrical features dictated by other design requirements.

4.5.1 Dielectric recovery of the remnant arc column

The ultimate limit to the dielectric recovery under the most ideal operating conditions is determined by the rate of recovery of the remnant hot channel following arcing at high currents. The dielectric breakdown of this remnant hot arc column is governed by a number of complex phenomena which have been studied in detail by Schade and Ragaller (1982), Ragalleer et al. (1982) Brand et al. (1982), Kopainsky (1978). For two metallic nozzles used in a duoflow configuration, the dielectric recovery follows the thermal recovery through three distinct phases (Fig. 4.7a). The first phase corresponds to a rapid recovery (2, Fig. 4.7a) followed by a second slower recovery phase (3, Fig. 4.7a). The final recovery phase (4, Fig. 4.7a) is again rapid.

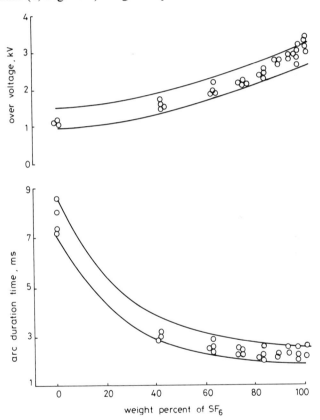

Fig. 4.6 a Performance as a function of SF_6 concentration (Fig. 8, Naganawa et al., 1985)

The dielectric recovery during each of these phases is governed by the temperature collapse of the remnant arc column (Fig. 4.7b). Immediately after current zero the temperature collapse is most pronounced at the nozzle throat and downstream. Initially the upstream region acts as an energy source for the downstream zones. Subsequently, the temperature of the upstream stagnant region decays more rapidly owing to the onset of turbulence, so that the temperature difference between upstream and downstream regions is significantly reduced. These conditions, which correspond to the first rapid recovery phase (2, Fig. 4.7a) imply the existence of positive and negative ions (but not so much electrons owing to their rapid decay) in the stagnation region. The charged particles, mainly atomic fluorine and sulphur, decay with a delay compared with the expected equilibrium concentrations (Fig. 4.7c) as already indicated in Chapter 2. Under the influence of an applied electric field, these residual charged particles move according to their mobility and form space charges which distort the electric-field distribution and hence the dielectric recovery (Moll and Schade 1979, 1980). Thus the field at the stagnation point is reduced at the expense of a corresponding increase downstream. The change from the space-charge-free field to the distorted field distribution (Fig. 4.7a) occurs in a finite time, so that the influence upon dielectric recovery depends on how rapid is the rate of rise of the transient recovery voltage. The dielectric strength during the first phase is therefore relatively low owing to the hot stagnant region upstream and the presence of atomic sulphur and fluorine, but it recovers rapidly owing to the rapid axial temperature decay.

The second, slower recovery phase (3, Fig. 4.7a) occurs due to a delay in the recombination process to form SF_6 from its components on the column axis.

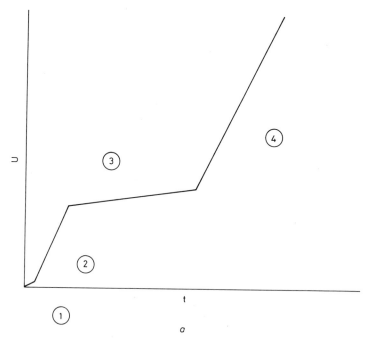

Fig. 4.7 *Dielectric-recovery characteristics*
a Shape of the recovery characteristic (Fig. 4, Schade, 1985)

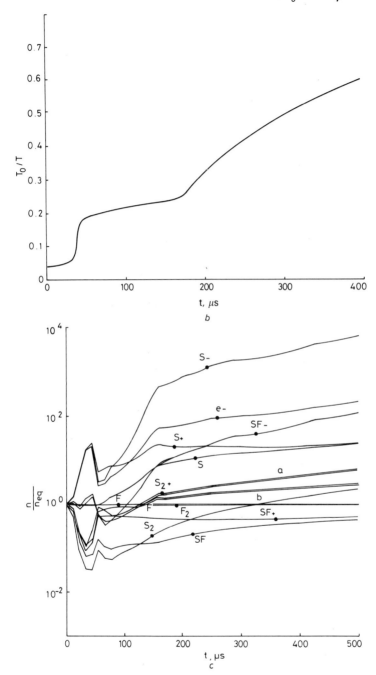

Fig. 4.7 *(continued)*
b Temperature variation during dielectric recovery (Fig. 4, Brand et al., *1982. Copyright IEEE)*
c Expected equilibrium concentration during dielectric recovery (Fig. 8, Kopainsky, 1978. Copyright Plenum Press)

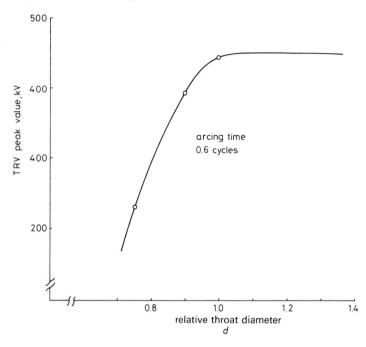

Fig. 4.7 *(continued)*
  d Effect of throat diameter (Fig. 8, Yanabu et al., 1981. Copyright IEEE)

The recombination heat is sufficient to maintain the axis at a temperature of about 1500 K for a prolonged period. Space-charge effects only persist during this phase of the recovery if a sufficiently steep recovery voltage occurred immediately after current zero. The breakdown strength during this phase is determined by whether a streamer is formed on the basis that the electric field becomes critical at least at one location along the hot-gas channel.

The third and final recovery phase (4, Fig. 4.7a) is characterised as being space-charge free and in thermodynamic equilibrium, so that the streamer criterion is easily applied to determine the breakdown voltage.

4.5.2 Scaling of the dielectric recovery of the arc column
Since all three phases of the dielectric recovery are governed by the temperature decay of the arc channel (Figs. 4.7a and b), the influence of different operating conditions upon the dielectric recovery are manifest through their effect upon the arc-column temperature. As a result, the magnitude of the voltage associated with, and the duration of each phase of, the recovery will, in general, vary with operating conditions. Consequently the establishment of scaling laws to take account of such influences is important for interrupter design studies.

Nozzle-geometry effects may be normalised by first scaling the voltage as the ratio of the electric field strengths at the stagnation point, and then scaling the

Characteristics of SF_6 interrupters 59

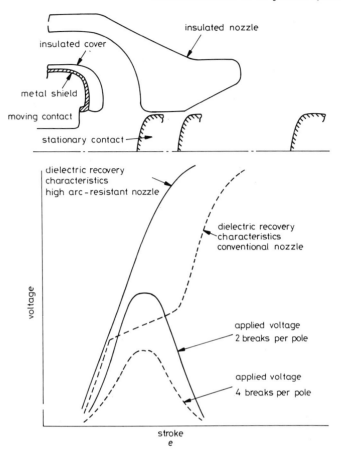

Fig. 4.7 *(continued)*
e Effect of new nozzle material (Fig. 6, Saito and Honda, 1982. Copyright Hitachi)

time according to theory (Schade, 1985). If the radial inflow into the volume between the two nozzles of a duoflow system is artificially made turbulent, the temperature decay in the stagnation region is accelerated, so leading to a more rapid recovery of dielectric strength. However, the final recovery to the non-arcing value is delayed for several milliseconds. The scale factors for non-turbulent-flow injection also apply for the turbulent-injection case.

Normalisation with respect to the upstream pressure may be achieved by scaling through the ratio of electric field strength to the particle density E/N at the stagnation point, but with unchanged time scales. Theoretically (Schade, 1985), a 20% reduction in the time scale is predicted when the pressure is doubled, but this has not been observed experimentally.

With pressure differences of the order of 0·1 bar across the nozzle, the duration of the thermal-recovery regime is increased by an order of magnitude. By contrast, the duration of the three phases of the dielectric recovery is only prolonged by a factor of 2 (Schade, 1985).

The fact that all three phases of the dielectric recovery scale in an identical manner, despite the governing physical processes being different, is attributed to the dominant effect of temperature upon these different physical processes.

4.5.3 Retarded dielectric recovery

The dielectric-recovery characteristics described above apply for a fully opened contact gap in which the contact effects have been minimised. In practice, the dielectric-recovery phase may coincide with the contacts being positioned at any location during contact travel. This leads to additional factors having a more dominant influence upon the dielectric recovery. Such factors include the length of the contact gap, the trapping of arc-heated gas in the nozzle entry, the influence of the nozzle shape and material, and the effect of any deterioration in contact condition owing to repeated arcing.

Tests performed with different contact-travel velocities (up to $5\,\text{ms}^{-1}$), so that the dielectric-recovery phase occurs at different contact positions within the main nozzle indicate that, for a sufficiently long stroke, the first two phases of the dielectric recovery (2 and 3, Fig. 4.7a) are still identifiable (Nagagawa et al, 1985).

However, if the stroke length is insufficient to provide a gas-flow area around the contact passing through the nozzle, which is twice the nozzle-throat area, then the first rapid phase of the recovery (2, Fig. 4.7a) vanishes and the gap recovery is consequently retarded. As a result, such effects determine the minimum arcing time for successful interruption. Furthermore, the breakdown voltage for a given stroke length decreases with peak fault current, although the puffer-chamber-pressure rise increases. Of course, without such pressurisation the reduction in dielectric recovery with increasing current could well be greater, as indeed appears to be the case with the thermal recovery (Section 4.4).

With shorter contact strokes and with higher arcing currents, gas heated during the high-current arcing phase may be trapped for long periods in regions of high electric field strength close to the nozzles and contacts. These regions of high dielectric stress may not, in practice, be completely eliminated on account of conflicting design requirements imposed by the puffer principle. Thus Yanabu *et al.* (1981) and Ikeda *et al.* (1984) have shown that hot gas trapped downstream between the fixed contact of a EHV interrupter and the nozzle exit (as a result of the required puffer action) can lead to the breakdown path being at the periphery and to the nozzle, rather than along the remnant hot-arc channel linking the upstream stagnation region, as considered by Ragaller *et al.* (1982). This is claimed to be caused by the considerably higher pressure (1–1·5 Mpa) which occurs upstream in a commercial interrupter compared with that used (0·36 Mpa) in the model interrupter tests (Ikeda *et al.*, 1984). As a result, the peak transient recovery voltage for a commercial EHV breaker which can be withstood decreases with short-circuit current, (Yanabu et al, 1981).

Characteristics of SF_6 interrupters 61

For quasi-steady conditions in SF_6 gas heated up to 2300 K, the breakdown voltage approximately obeys Paschen's law (e.g. Eliasson and Schade, 1977; Nagata et al., 1980). There is evidence that the breakdown is dominated by the surface field strength close to the cathode (e.g. Nitta et al., 1975), so that the voltage which causes breakdown at a mass density ρ is given by (Kuwahara, 1983):

$$V_{crit} = \frac{E_{0crit}}{E_{cathode}} \frac{\rho}{\rho_0} V_a \tag{4.9}$$

where E_{0crit} is the critical field strength at a mass density ρ_0 and $E_{cathode}$ is the electric field on the cathode when a reference voltage V_a is applied across the gap.

Thus the lower gas density which persists in the region close to the fixed contact with smaller-diameter nozzles and also with prolonged arc duration leads to inferior dielectric recovery in accordance with eqn. 4.9. Improved dielectric recovery can be achieved not only with a larger nozzle throat (Fig. 4.7d), but also with the use of a Bell-shaped downstream section (Fig. 4.5a) as opposed to a conical section, probably on account of improved ionic recombination (Section 4.3.2) and more directed exhaust of hot gases.

The downstream breakdown identified by Yanabu et al. (1981) and Ikeda et al. (1984) appears to involve surface breakdown along the PTFE nozzle. This, in turn, may well involve conduction through the partial carbonisation of the PTFE produced by frequent high-current interruptions (Saito and Honsa, 1982). As a result, there may occur a premature pause in the dielectric-recovery characteristics of the type measured by Nakagawa et al. (1985) until the contact has cleared the defective nozzle region. However Saito and Honda (1982) claim that this deficiency may be overcome by using a new arc-resistant material which suffers less carbonisation (Fig. 4.7e).

Account also needs to be taken of aging effects produced by frequent high-current interruption, which may increase the roughness of the contact surface. Electrostatic shields (Fig. 4.7e) may be profitably utilised to reduce the electric field strength at the contacts below the critical value dictated by eqn 4.9.

4.5.4 Dielectric-recovery requirements for medium-voltage interrupters

The rapid initial dielectric-recovery phase observed by Ragaller et al. (1982) for the decaying arc channel is such that a voltage withstand of 100 kV may be achieved typically in 50 μs (i.e. 2 kV/μs). This approximate recovery rate is, of course, of the correct order to meet IEC specifications for medium/low-voltage interrupters used for distribution-system applications. The implications are therefore that dielectric-recovery requirements for such interrupters should be comfortably met by SF_6. Indeed, tests with overstressed rotary-arc interrupters, in which the remnant arc channel might be expected to persist longer than in a puffer-type interrupter, indicate less susceptibility to dielectric rather than thermal failure.

4.5.5 Gases with higher dielectric strength

The quest for gases with better dielectric strength than SF_6 has been pursued by many researchers (e.g. Christophorou, 1980), but the problem has been placed in perspective by Brand (1982). Brand finds that the breakdown field strength, as well as the boiling point and toxicity, of a gas is related to the dipole polarisability and the ionisation potential of the molecules. However, gases with the best dielectric strength E_r are not the ones with the highest ionisation potential E_i, because the dependence upon the polarisability α is stronger, i.e.

$$E_r = 5.426 \times 10^{-3} E_i \alpha^{1.5} \tag{4.10}$$

Examination of a range of possible candidate molecules indicates that the absolute limit of $E_r = 6$ (Brand and Kopainsky, 1979) cannot be reached, and that there seems little hope of finding unitary gases with more than three times the dielectric strength of SF_6. Gases possessing good insulating qualities like SF_6 ($E_r \sim 0.8 - 1.0$) are exceptions compared with the mean, on account of a special molecular structure.

Chapter 5
Arc modelling and computer-aided methods for interrupter-design evaluation

The proliferation of interrupter types, based not only upon a given arc-control and quenching method but also upon different operating principles, has highlighted the need for computer-aided-design packages to assist during initial design evaluation. The possibility of also being able to provide guidance to prospective clients concerning the predicted performance of the circuit breaker with respect to particular conditions which might prevail in the clients network should also not be overlooked, and could be particularly attractive for isolated 'stand alone' networks with unique characteristics. The usefulness of computer-aided-design packages is therefore clear, and has been recognised through the formation of a CIGRE Working Group concerned with arc modelling, whilst both industry and universities worldwide are making progress in establishing practically useful modelling and computational packages.

Such models and computer packages concern one or all of the major functions of a circuit breaker (e.g. China, 1984), namely conducting and interrupting the current, providing adequate insulation levels and ensuring sufficient mechanical drive for efficient operation. The interruption function may be subdivided into a number of secondary functions which correspond to particular components of the circuit breaker, as indicated on Fig. 5.1. The interdependence of these many components upon particular functions reveals the difficulty of producing a completely computerised design procedure. Thus it may be possible to establish particular relationships to represent some of the functional links which lead to the dimensioning of a given component, but it is impossible to perform these operations for several such links. Consequently, it does not seem feasible at present to design a circuit breaker completely by such means, but nonetheless parts of the design procedure may be undertaken more efficiently using computer-aided-design methods. The creation of such computer-aided-design packages needs, as basic elements, accurate arc modelling, as well as the prediction of aerodynamic, thermal, electrostatic and electromagnetic fields and the prediction of the mechanical behaviour of the contact-drive system.

If, on the other hand, computer-based predictions of the interaction of the

64 Interupter-design evaluation

circuit breaker with particular electrical networks are required, it may be more appropriate to regard the arc discharge as the coupling between the network and the aerodynamics of the circuit breaker. Consequently, the approach suggested by Leclerc et al. (1980) becomes attractive, whereby the aerodynamic system is modelled with an electrical analogue so that the entire problem is transformed into the description of two electrical networks interconnected through the arc discharge.

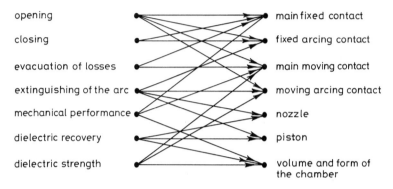

Fig. 5.1 *Dependence of functions upon various parts (Fig. 3, China, 1984. Copyright CIGRE)*

5.1 Field calculations

A variety of techniques are available for calculating time and space variations of field (e.g. Tandon et al., 1983; Armstrong and Biddlecombe, 1982; Mayer and Nagamatsu, 1980; Ryan et al. 1983; Scott et al., 1974). Provided that the geometry is symmetrical, the field may be discretised using either finite-difference or finite-element techniques, or, in the case of electromagnetic fields, a coupled-circuit approach may be employed (e.g. Turner and Cheng, 1985). Difficulties with finite-element methods are associated with the interdependence of parts, but these may be overcome using automatic or semi-automatic mesh-generation programs. For 3-dimensional geometries without symmetry, use is made of the boundary integral-equations method (e.g. China, 1984).

Some typical examples of field calculations reported by several authors are shown in Fig. 5.2. Nagamatsu et al. (1983) have studied axisymmetric incompressible flow in duoflow nozzles using a finite-element (first-order triangular elements) program with an average of 10^3 elements per geometry. Fig. 5.2a compares the predicted non-arcing flow in two duoflow nozzles having different inlet-wall radii and gap spacings. The flow fields are shown as Mach-number contours, from which it may be observed that the flow which dominates thermal recovery close to the nozzle axis is similar for both nozzle geometries. Studies of this kind have shown (Noeske et al., 1983) that, for a fixed inlet-wall radius, the axial-flow acceleration is constant over a range of nozzle spacings but limited for small gaps by choking of the inlet and for large gaps by flow instabilities. Furthermore, the size of the central stagnation region, and therefore the susceptibility of the nozzle to degradation by trapped warm gas, is greater for a larger nozzle-gap spacing and/or for a greater inlet wall radius.

Similar field calculations have been used by Shimmin (private communication) to study the manner in which the non-arcing flow field in a partial duoflow interrupter varies with contact separation during the contact-opening process.

Flow-field calculations with arcing have not been so extensively studied, but recently Fang and Newlands (1984) have reported how the flow through a nozzle is modified by arcing at currents up to about 60 Ka (Fig. 5.2b). Their results show that, whereas in the absence of arcing a wake exists downstream of the upstream contact, with an arc present there is flow entrainment into the arc column as a result of the rapid acceleration arising from the increased sonic velocity of the hot gas. At higher currents the interaction of the flow with electromagnetic pinch forces close to the upstream contact produces vortex flows at the arc periphery (Fig. 5.2b).

Ragaller et al. (1982) have utilised a transformation of the conservation equations for mass, momentum and energy into isotherms for providing a thermal map of a decaying current-zero arc column in a duoflow-nozzle system. Initially 25 discrete isotherms are utilised. A typical example of the decaying thermal field is shown in Fig. 5.2c corresponding to 30 and 40 μs after current zero. From such an analysis, the importance of turbulence at the stagnation

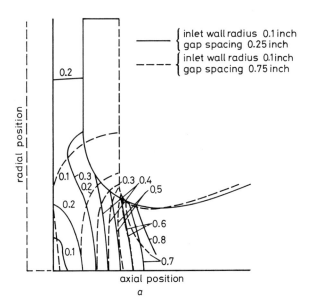

Fig. 5.2 *Field calculations*
  a Comparison of single- and double-nozzle flow Mach numbers (Fig. 2.2, Noeske et al., 1983)

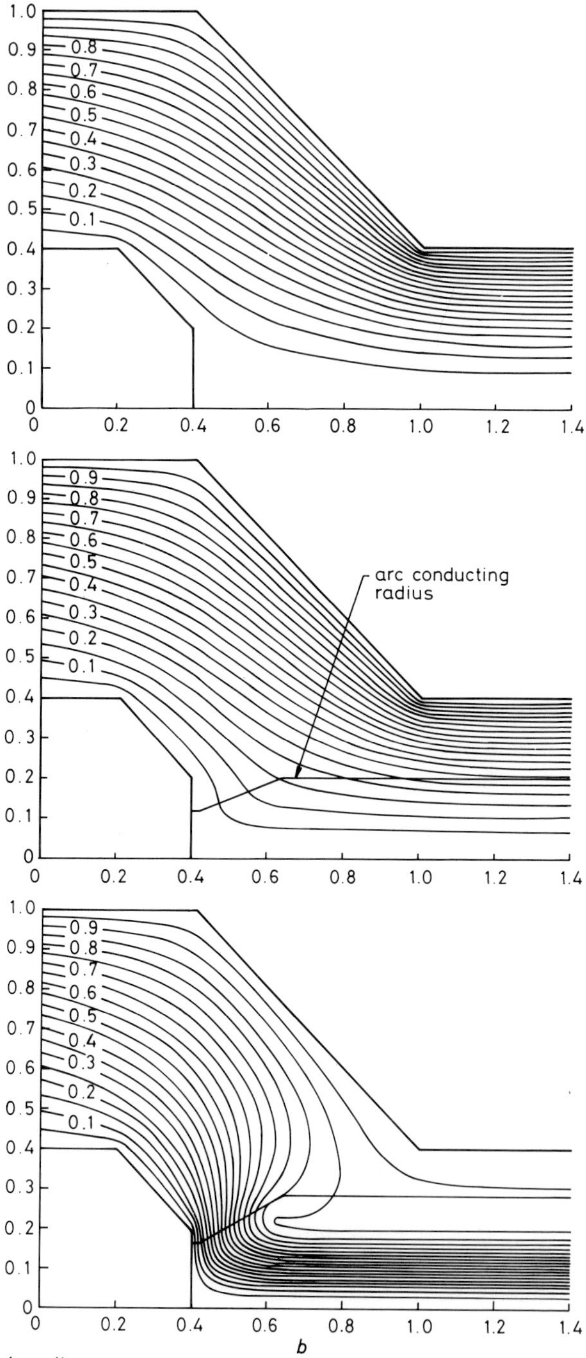

66 Interupter-design evaluation

Fig. 5.2 *(continued)*
b Effect of arcing on flow field (Fang and Newland, 1984)

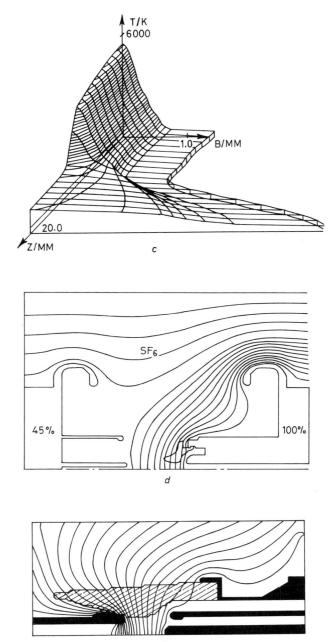

Fig. 5.2 *(continued)*
  c Effect of arc on thermal field (Ragaller et al., 1982. Copyright IEEE)
  d Effect of polarity on electrostatic field (Fig. 7, Ryan et al., 1983)
  e Effect of contact movement on electrostatic field (Fig. J-8, Noeske et al., 1983)

68 Interupter-design evaluation

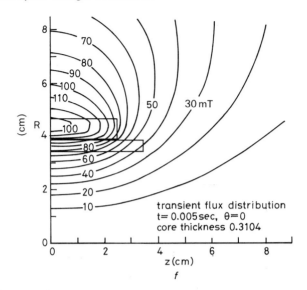

Fig. 5.2 (continued)
f Electromagnetic field in rotary-arc interrupter (Chen and Turner, private communication)

region and deviations from local thermal equilibrium have been identified during the dielectric-recovery period.

Several examples of electrostatic-field calculations directed at identifying insulation weaknesses are available in the scientific literature (e.g. Noeske et al., 1983; Ali and Headley, 1984; Ryan et al., 1983; Yanabu et al., 1981). Fig. 5.2d shows the field distribution in the vicinity of the higher-voltage break of a 2-break 420 kV 63 kA 2-cycle-operation puffer circuit breaker (Ryan et al., 1983). The field distributions were derived from the predicted fields for a 4-break interrupter using established empirical techniques, and show the manner in which the field distribution depends upon the polarity of the restrike voltage. Such calculations assisted in the detailed profiling of the shields to alleviate excessive fields in the more susceptible regions close to the contacts, where gas could be trapped and degrade the insulation strength. Furthermore, the manner in which the field distribution is modified in the vicinity of the grading capacitor may also be calculated. Fig. 5.2e shows the excessive fields produced around the tip of the fixed contact during the movement of a nozzle with narrow downstream diverging section, thus suggesting that a bell-shaped divergence could be advantageous (Noeske et al., 1983).

Electromagnetic-field calculations have proved particularly useful in the design of rotary-arc interrupters. Turner and Cheng (1985) have shown that good agreement exists between the predictions of a finite-difference solution for the magnetic vector potential (together with a Crank–Nicolson solution of the transient problem) and results obtained with a coupled-circuit model of the system. Fig. 5.2f shows the field distribution in a quadrant of the South Wales Switchgear type of rotary-arc interrupter (Turner and Cheng, private communication), from which details of important magnetic wells have been identified.

Results for the Merlin-Gerin type of rotary-arc interrupter (China, 1984) show that different magnetic-field distributions occur in a compact breaker housing when only the central phase is concerned and when all three phases are taken into account.

5.2 Arc modelling

Arc modelling is essentially concerned with predicting the electrical response of the arc through either an explicit or implicit knowledge of temporal thermal changes in the arc column. Rigorously, this may, in principle, be achieved through the simultaneous solution of the differential equations (involving time, radial and axial differentials) expressing mass, momentum and energy conservation along with Ohm's law and Maxwell's equations. Even with modern computational facilities such an approach can only be applied to a limited number of situations, since the defining equations are highly nonlinear and some of the complicated discharge conditions such as radiation transport (particularly for a contaminated plasma) are difficult to define mathematically. Consequently, arc modelling still involves the need to make some simplifications, which unfortunately are often not stated explicitly.

The simplest approximation is to reduce the problem to a single ordinary differential equation with respect to time, the most well known examples of which are the Cassie, Mayr and Browne models (e.g. Flurscheim, 1982). Physically the Cassie model assumes that the dominant loss of energy is by axial convection, whereas the Mayr model assumes that radial diffusion at the arc boundary dominates. Unfortunately, these models do not take sufficiently detailed account of the physical processes occurring in the arc column, so that in many cases serious discrepancies between predictions and test results are observed. Nonetheless attempts to incorporate non-thermal-equilibrium effects during the recovery period have been made by Reider and Urbanek (1966).

At the other extreme, Ragaller *et al.* (1982) have attempted a full 3-dimensional (radius, axis, time) solution of the conservation equations for the particular case of the post-current-zero period, but neglect radiative transport and make assumptions about the role of turbulence. The rigour of their approach is therefore limited, although computational effort is considerably increased. Nonetheless, problems associated with non-thermal equilibrium have been taken into account.

In between these two extreme approaches there are a further two classes of models in which either the axial or radial dependency is simplified. Reasonable, if limited, predictions have been made by neglecting any axial dependence and making assumptions about the axial location which dominates the arc-gap behaviour (e.g. Hermann and Ragaller, 1977).

70   Interupter-design evaluation

Unfortunately, there are important axially dependent phenomena such as flow straining (Fig. 5.3) (e.g. Jones, 1977), which cannot be incorporated when axial variations are neglected. A better approach involves retaining the axial variation (e.g. Fang, 1983; Swanson, 1977; Tuma and Lowke, 1975; Tuma, 1980; Swanson and Roidt, 1971; Chan et al., 1976, 1978). Each model may be regarded as a special case of the rigorously formulated boundary-layer integral analysis of Cowley (1974).

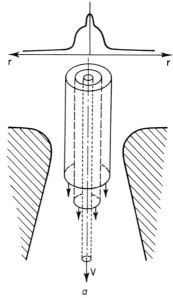

Fig. 5.3 *Arc modelling-flow straining effects (Fig. 4, Kopainsky, 1978. Copyright Plenum Press)*

The boundary-layer integral method describes arc behaviour on the assumption of similarities between the radially integrated properties of various arc types. Thus a detailed knowledge of the radial arc structure becomes redundant once the scaling laws relating the various radial integrals have been established. The method also enables formal arc analysis to be blended with empiricism, so that discharge conditions which are difficult to define mathematically (e.g. radiation due to electrode-vapour contamination) may be accomodated quantitatively. Turbulence effects are automatically incorporated within the radial integrations.

The various radial integrals are expressed as 'characteristic areas'. One of the simplest characteristic areas to comprehend physically is the conductance area θ_c, which represents a normalised electrical conductance of the arc, i.e. Electrical Conductance:

$$\theta_c = \sigma^* \int_0^\infty (\sigma/\sigma^*) 2\pi r \, dr \qquad (5.1)$$

where σ^* is a reference value of electrical conductivity. Another physically important area is the thermal area $\theta_\delta = \int_0^\infty (1 - \rho/\rho_\infty) 2\pi r dr$ (ρ_∞ is the cold-flow mass density) due to arc heating. Physically, the difference between the nozzle cross-section and the thermal area $(A - \theta_\delta)$ gives the effective flow cross-section of the nozzle-arc combination. Thus the characteristic areas take account of the magnitudes, shape and radial extents of the radial profiles. Closure of the defining equations is achieved through shape factors Λ, which relate the various areas to the thermal area, i.e. $\Lambda_c = \theta_c/\theta_\delta$. The shape factors reflect the influence of the shapes and magnitudes (but not radial extent) of the temperature and velocity profiles. Finally, the shape factors are interrelated through other characteristic quantities such as arc dynamic power loss, the relationships being determined from either test tesults (Walmsley and Jones, 1980) or numerical simulation (Chan et al., 1976).

Investigations show that unique relationships exist between the various shape factors and the arc dynamic power loss. For high-current arcs, all shape factors are approximately constant and radiation losses may be incorporated empirically (Strachan et al., 1977). The integral analysis then predicts the current density for thermal blocking of a nozzle (Fang et al., 1980):

$$j_t = \frac{p_0 \sqrt{h_0 \sigma^* k}}{z_t N_b} \frac{\Lambda_h \Lambda_c}{\Lambda_d^2} \qquad (5.2)$$

where p_0 is the upstream pressure, h_0 is the stagnation enthalpy, k, $\Lambda_{hl} \Lambda_c \Lambda_d$ are enthalpy, conductance and displacement shape factors, z_t is the upstream arc contact gap and N_b is the nozzle coefficient uniquely determined by the nozzle geometry. The integral analysis can also accomodate nozzle-ablation effects by introducing an additional non-dimensional coefficient which is determined by the thermodynamic properties of the nozzle material, the characteristic quantities of the gas and the radiation coefficient for a given upstream electrode material.

During the arc-recovery period, the method predicts satisfactorily the thermal-recovery behaviour, which can be used to investigate the influence of the interconnected electrical network (Tahir, 1978) and the influence of different nozzle shapes (Fang and Brannen, 1979). The approach may therefore be used for initial design evaluation of the nozzle system with regard to the thermal-recovery and nozzle-blocking limits (Fang and Jones, 1979).

The difference between the various radial integral models lies in the manner in which the integral equations are closed (Fang, 1983). Thus, whereas the rigorous analysis utilises shape factor and dynamic power-loss relationships, the analysis of Swanson and Roidt (1972) assumes a less realistic Bessel function for the radial profile. The Brown Boveri Group (e.g. Hermann and Ragaller, 1977) assume radial temperature and velocity profiles of the same shape, but of variable magnitude and radial extent, and their model requires the input of six parameters, the values of which cannot be determined by a consideration of the physics or dimensions of the system. The model used by Tuma and Lowke (1975), Lowke and Ludwig (1975) and El-Akkari and Tuma (1977) assumes that

72 Interupter-design evaluation

the temperature and axial velocity are uniform across the electrically conducting channel. With appropriate assumptions about the axial and radial variables, it may be shown that the models of Cassie and Mayr are also special cases of the integral-boundary-layer analysis (e.g. Swanson, Roidt and Browne, 1972; Shimmin, Private communications). Of course, each of the above models involves additional approximations, as detailed by Fang (1983).

5.3 Puffer modelling

Several models of puffer-type interrupters have been described in the scientific literature (e.g. Murai et al., 1981; Ueda et al., 1982; Ueda et al., 1979; Yanabu et al., 1981). Most of these models deal with the complicated compressions which occur in, for instance, the piston, arc and suction chambers (e.g. Veda et al, 1982), but often at the expense of arc modelling (e.g. neglect of such important processes as radiation loss and ablation) and the extent to which heated and cool gases mix in the arc and piston chambers. These aspects have been considered further by Shimmin (1986).

The arc-chamber pressure is determined by the combined actions of piston compression, nozzle blocking and arc heating, as indicated by the mass and energy exchanges between the various volumes shown schematically in Fig. 5.4. In the absence of arcing, the pressure rise in the arc chamber is governed by the balance between the compressed-gas volume and volume flowing through the nozzle, and the compression rate (Yanabu et al., 1981). The simplest form of this expression is

$$p = p_0 \{\exp\left[((A_A a/2C_C V) - 1)\ln(l_0 - Vt)/l_0)\right]\}^{\delta_0} \tag{5.3}$$

where A_A, C_c are the cross-sectional areas of the nozzle throat and puffer cylinder, respectively, a is the sonic velocity, V is the piston speed, l_0 is the closed length of the puffer cylinder, t is time and δ_0 is the specific-heat ratio of SF_6 at normal temperature. In the presence of arcing, which merely causes nozzle blocking but no excess heating of the arc-chamber gas, the nozzle area is replaced by the effective flow area $(A_A - \theta_\delta)$ (Section 5.2), where θ_δ is the thermal area of the arc and may be determined from the boundary-layer integral-arc model. There is evidence that this situation probably prevails in some commercial duoflow breakers. The empirical relationship given by Sasao et al. (1981) (Fig. 4.2b), which relates electrical-energy input to the size of the thermal volume of arc-heated gas, may be used to calculate the pressure rise (Fig. 4.2c). Alternatively, for complete gas mixing the pressure rise can be calculated by the gas law.In all cases, the influence of mass injection due to nozzle ablation needs to be incorporated. This situation exists at least is some commercial partial duoflow puffer interrupters.

The extent to which arc heating may be tolerated to assist in providing the pressure rise (with the consequent advantage of a reduction in piston-chamber

Interrupter-design evaluation

volume) is governed by dielectric-breakdown limitations to performance, which are in turn determined by breaker-terminal faults (Chapter 9).

The opening speed V of the interrupter contacts is approximately determined by the required interrupting time, and may be adjusted to obtain the required pressure. The length of the puffer cylinder is determined by the withstand-voltage requirements. The value of $(l_0 - V_t)/l_0$ that gives the maximum pressure is determined by the dead volume and the deceleration of the opening speed at the termination of opening. Since these considerations determine P, A_A, V, l_0 and the contact position at which the pressure rise is stopped, the required cross-section C_c of the piston chamber may be obtained from eqn. 5.3.

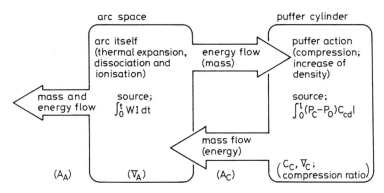

Fig. 5.4 *Puffer modelling: Energy and mass exchange in a puffer geometry (Fig. 4, Ueda et al. 1982. Copyright IEEE)*

In the case of partial duoflow interrupters, the cross-section of the secondary nozzle may be determined from the ratio $A_{A2}//A_{A1}$, which depends upon the diameters of the main nozzle, fixed-arc contact and the stroke of the moving-arc contact. For low-capacity circuit breakers (i.e. distribution-type breakers), A_{A2}/A_{A1} is typically 0·5 to 0·8 (Yanabu et al., 1981), but for high-capacity breakers (i.e. EHV breakers) the value will be greater.

Typical results derived from the above type of design process (Yanabu et al., 1981) give the ratio of puffer-cylinder volume as a function of the interrupting capacity VI with the required recovery voltage as parameter.

5.4 Rotary-arc-interrupter modelling

The modelling of rotating-arc columns has been more concerned with arc-heater applications (e.g. Chen and Lawton, 1968; Schrade, 1973) rather than circuit interrupters, and as such involve quasi-steady rather than transient calculations. However, Spencer et al. (1985) have recently developed an approximate computer model for predicting the arc behaviour and thermal-recovery performance of a rotary-arc interrupter.

It has been shown that, unlike the axial-flow arc, the rotary-arc plasma column has approximately uniform electrical properties along its length, apart from the contact regions (Spencer *et al.*, 1985). Consequently the axial variables in the arc-conservation equations may be ignored, and the arc can be modelled on the basis of cross-flow considerations (e.g. Uhlenbusch, 1976) during the high-current arcing phase.

A knowledge of the magnetic-field distribution produced by the fault current allows the arc shape and velocity to be determined from the balance of Lorentz driving force, drag-retarding force and arc-momentum change. The arc behaviour during the high-current phase is governed by convective and radiative

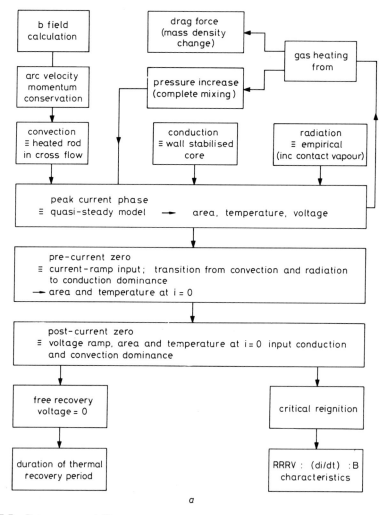

Fig. 5.5 *Rotary-arc modelling*
 a Rotary-arc thermal-recovery caclulation scheme

Interrupter-design evaluation

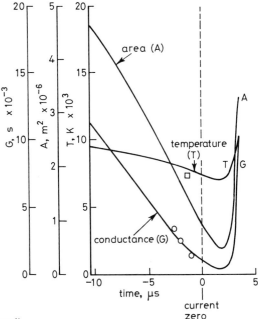

Fig. 5.5 *(continued)*
b Comparison of experiment and theoretical predictions of rotary-arc recovery (Fig. 7, Spencer et al., 1985)

losses, the former being determined from a knowledge of the arc velocity and the latter from empirical relationships taking account of electrode-vapour entrainment (e.g. Jones and Fang, 1980). The arc voltage determined from this calculation enables the gas heating to be evaluated (Fig. 5.5a), and hence the change in mass density and pressure. The former affects the arc velocity, which is upgraded, and the latter affects the radiative emission, which is also adjusted.

The pre-current-zero period is modelled with a current-ramp input, a transition from convection to conduction dominance is identified and the arc-column cross-section and temperature at current-zero are determined. These latter parameters, along with a voltage ramp, form the input for the post-current-zero period, during which the relative importance of convective and conduction losses depends upon the phase of the driving magnetic flux density and the magnitude of the post-arc current. For no post-current-zero voltage input, the resulting free recovery calculation allows the approximate duration of the thermal recovery period to be determined. These estimates are in agreement with test results (Spencer, private communication). With a post-current-zero voltage ramp, the value yielding critical reignition is determined, thus enabling the RRRV versus di/dt characteristic to be established with magnetic flux density as parameter.

Some typical predictions of arc behaviour during the current-zero period are compared with experiment in Fig. 5.5b.

Chapter 6
Impact of SF_6 technology upon transmission switchgear

The present trend is for transmission circuit breakers to be of the SF_6 puffer type. The historical evolution of such interrupters can be traced, along two major strands which correspond respectively to the development of SF_6-based technology and the puffer principle.

Early development of the puffer interrupter occurred in both Europe and the USA, Delle of France having a load breaker using air as the arc-quenching medium in the 1930s. However, because of the inadequacy of air for puffer action (Section 4.1), the puffer concept remained dormant until 1960 when Westinghouse introduced a series of SF_6 puffer breakers for voltages of 34·5 to 69 kV. Even so, Westinghouse and Siemens thereafter preferred to concentrate upon 2-pressure SF_6 breakers as being more promising for the higher voltage ratings.

At this stage, 420 kV 2-pressure circuit breakers were being employed in the UK, utilising air as the quenching medium and with 12 breaks in series per phase to achieve the required performance. At the same time, SF_6-insulated current transformers were produced over the range 146–420 kV. The first transmission SF_6 circuit breakers in the UK at 145 kV were manufactured by GEC and commissioned in 1966. Siemens had already produced 123 and 145 kV duoflow 2-pressure breakers in 1964 (Beier *et al.*, 1981). These followed air-blast technology in being 2-pressure interrupters, but with a totally enclosed system.

The first transmission SF_6 puffer circuit breakers in continental Europe were introduced by Magrini and Delle, and in Japan by Mitsubishi in 1965. In the UK, GEC and in continental Europe, Siemens abandoned the 2-pressure design for the puffer in 1974 (Beier *et al.*, 1981). Currently GEC and NEI (Reyrolle) Ltd., both produce puffer type SF_6 circuit breakers.

The growth in the number of dead-tank-type SF_6 circuit breakers for various voltage ratings in service in Japan in 1974–81 is shown in Fig. 6.1a (Nakanishi *et al.*, 1982). By the end of 1981 Mitsubishi claimed to have 4385 SF_6 circuit breakers in service (Kuwahara *et al.*, 1983a). Hitachi, who introduced SF_6 circuit breakers in 1969, now manufacture SF_6 circuit breakers in the voltage range 72–550 kV with a rated interrupting current of 20–63 kA (nominal current 1·2–12 kA), and claim to have 3500 working satisfactorily in service.

Impact of SF_6 technology upon transmission switchgear

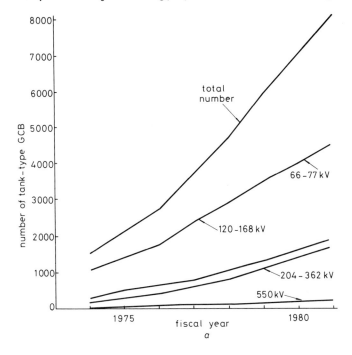

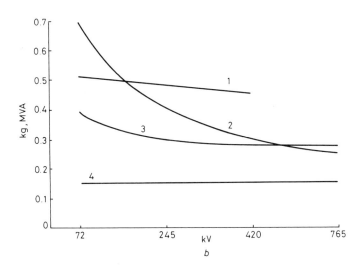

Fig. 6.1 *Trends in SF_6 transmission circuit breakers*
a Increase in the number of tank-type circuit breakers in Japan (Fig. 1, Naķamishi et al., 1982. Copyright CIGRE)
b Ratio of weight to power for outdoor circuit breakers (Fig. 1, Eidinger and Petitpierre, 1979. Copyright Brown Boveri)

78 Impact of SF_6 technology upon transmission switchgear

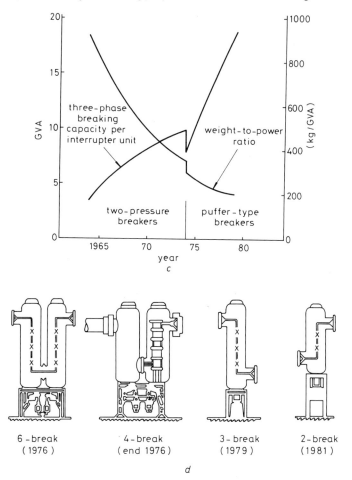

Fig. 6.1 *(continued)*
c Breaking capacity as a function of weight (Fig. 1, Beier et al., 1981)
d Stages in the evolutionary development of a dead-tank circuit breaker (Fig. 4, Ali et al., 1982 and Fig. 1, Ali and Headley, 1984)

GEC (UK) have installed open-terminal-type circuit breakers in the UK and in tropical climates throughout the Far East. Their 4-gap-type FE4 circuit breaker, which has been installed in the UK for 50 kA 420 kV duty, has withstood 2·9 MV across the contacts, representing voltage doubling of a standard 1·425 MV lighting impulse. Their 6-gap circuit breakers have been in service since 1978 and service experience exceeds 100 circuit-breaker years (Harris, 1984). Similarly, NEI Reyrolle (UK) have installed open-type circuit breakers in the UK and abroad covering the ranges 145 kV (1 break, 40 kA), 245/300 kV (2 break, 50 kA) and 420 kV (4 break, 50 kA). They have also recently developed a single-break puffer rated at 420 kV 63 kA.

Impact of SF_6 technology upon transmission switchgear 79

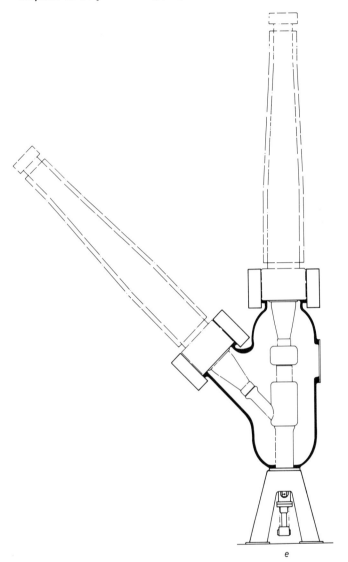

Fig. 6.1 *(continued)*
e 420 kV single-break arrangement

ASEA (Sweden) and BBC (Switzerland) produce SF_6 circuit breakers capable of operation under Arctic conditions at temperatures as low as $-50°C$; mixtures of SF_6 and nitrogen are utilised to overcome low-temperature liquefaction problems. Although the same voltage rating can be maintained as for pure SF_6, the fault-current rating is somewhat reduced (e.g. from 40 to 31·5 kA) (Solver, private communication).

80 Impact of SF_6 technology upon transmission switchgear

Currently circuit breakers for 550 kV transmission systems are available (e.g. BBC, Siemens, Toshiba) and development work on still higher voltage breakers up to 800 kV (e.g. BBC, Siemens) is well advanced. Concurrently SF_6 generator breakers of various designs are being made available (e.g. BBC, Mitsubishi) with fault-current ratings comparable to the more traditional air-blast generator breakers. There is also a trend to rely more on SF_6 interrupters, either alone or in tandem with vacuum interrupters, as the components of circuit breakers for high-voltage D.C. transmission systems.

6.1 Commercial considerations

The impact of developments in SF_6 technology upon circuit-breaker economics may be considered in terms of the weight per MVA capacity of the breaker instead of the cost, since, for minimum oil, air blast and SF_6, similar raw materials and manufacturing costs are involved (Eidinger and Petitpiere, 1979). (Bulk-oil circuit breakers on the other hand involve considerably higher raw-materials costs and lower manufacturing costs). In 1979, a comparison of the weight per MVA as a function of operating voltage (Fig. 6.1b) showed a clear economic advantage for SF_6 compared with minimum oil and air up to about 500 kV, and comparable costs to air blast above 500 kV. Since 1979, the further evolution of SF_6 technology (particularly with self-extinguishing principles) has increased the economic advantage of SF_6 throughout the entire voltage range (Fig. 6.1b curve 4; Eldinger, 1987, private communication). The reasons for the ascendancy of SF_6 circuit breakers are numerous and complexly interrelated. They are not solely due to the superior dielectric strength of SF_6, but also associated with the attractive compressive properties which are necessary for good puffer-interrupter performance (Chapter 2).

As already discussed in Chapter 2, the advantages of SF_6 over air in a 2-pressure interrupter owing to its faster recovery speed are partly compensated by some disadvantages. The arcing products of SF_6 are chemically more aggressive than those of air, resulting in the need for a higher selectivity of compatible materials. For low-temperature installations the tendency for SF_6 to liquefy necessitates the use of SF_6–N_2 mixtures (e.g. Bross, 1981). The lower recovery speed of air can at least be partly compensated for by the use of higher pressures. However, in the case of the single-pressure SF_6 breaker, the gas may be stored at a relatively low pressure because the interruption pressure is produced by compression in the puffer cylinder during breaker operation.

Despite these facts, 2-pressure SF_6 breakers were for a long time the only type available for transmission duties for a number of reasons. Manufacturers of 2-pressure SF_6 breakers could rely on well established experience with air-blast interrupters, and could rely upon earlier optimisation efforts (e.g. Noeske, 1983; Beier et al., 1981). With such breakers, the upstream receiver pressure can be assumed to be an independent parameter for optimisation as long as the nozzle is not clogged, so that the search for optimisation is simplified. With puffer breakers, pressure transients are induced and nozzle blocking during arcing is used for gas conservation, so that the parameters of the system dynamics are all interrelated in a complicated manner as discussed in Chapter 8. The nature of the quenching medium has a further important influence on pressurisation through heating and gas conservation governed by mass flow, as discussed in Chapter 4. Consequently, optimisation of a puffer-breaker design is a more complicated task, since, for example, the contact travel speed and piston movement can vary with interruption current if the activator is not over-designed (e.g. Chapter 8).

Although there is evidence from fundamental interrupter studies that several aspects of puffer optimisation remain to be explored (e.g. the voltage-withstand-limitations of self pressuring interrupter), in commercial terms the gains attainable with the puffer principle are impressive even with the present degree of optimisation. For instance, the puffer breaker in its simplest form becomes economically competitive with low-oil-content breakers because it requires no expensive valves. Furthermore, the single-pressure interrupter operating mechanism is less expensive and less complex than that of the 2-pressure breaker. The interrupting capacity per break of SF_6 interrupters has been increased by almost an order of magnitude during the last 15 years (Fig. 6.1c). Although this in part was achieved with 2-pressure interrupter technology, the significant impact made by the single-pressure (puffer) interrupter technology in the mid-1970s led to a further increase which is still progressing. This remarkable improvement has been achieved along with an equally dramatic decrease in the weight/capacity ratio of the interrupters (Fig. 6.1c). The gain achieved in this ratio due to the introduction of the puffer principle in the mid 1970s is again apparent (Fig. 6.1c). The increased breaking capacity results not only in a reduction in the weight/capacity ratio but also in the number of breaks required per phase to achieve a given performance. Thus, whereas 2 breaks per phase were required to meet a 300 kV, 50 kA rating with a 2-pressure SF_6 breaker just before the introduction of the puffer breaker, the same performance could be achieved in 1981 with a single break using a puffer breaker (Fig. 6.1c; Yanabu et al., 1985). The impact of the reduction in the number of breaks per phase is not only upon the circuit-breaker weight but also upon the complexity of the operating mehanism (which has to open fewer gaps), as well as a reduction in circuit-breaker size. The comparable size reduction which has been achieved is illustrated in Fig. 6.1d (Ali et al., 1982). This trend has continued since 1981, and NEI (Reyrolle) have recently developed a single-break puffer breaker rated at 420 kV, 63 kA (Fig. 6.1e), which is a considerable improvement over the 12 breaks utilised with air about 20 years ago.

Hitachi also give figures for the reduction in the number of components needed to construct a circuit breaker as a result of the introduction of SF_6 technology; for the 204–300 kV range, the number of parts has been reduced by 40–50%, whilst for a 550 kV breaker the reduction is somewhat less at 20–30%.

82 *Impact of SF_6 technology upon transmission switchgear*

Of course, by suitable design many common components can be used for a range of different circuit breakers. Consequently, better production control can be achieved, resulting in a higher circuit-breaker reliability.

6.2 Circuit-breaker assemblies

Most circuit-breaker manufacturers produce both live- and dead-tank versions. In live-tank designs the interrupters are housed in porcelain insulators isolated above ground level (Fig. 6.2a). The overall construction is similar to the older small-oil-volume circuit breaker. With the dead-tank construction the interrupters are housed in an earthed metal enclosure at ground level. Although live-tank breakers are generally considered to be the most economic for open-terminal (i.e. conventional connection in open air to the transmission system) applications, the dead-tank construction with bushings can sometimes result in overall economies. For instance, if current transformers are required on both sides of the circuit breaker, the cost of these components should be taken into account overall (Ali and Headley, 1984).

The range of live-tank circuit breakers shown in Fig. 6.2a, which covers the voltage levels of 145–800 kV is typical of the range offered by many manufacturers worldwide. Details of the interrupter unit, which is housed at the end of the vertical stalk, are shown in Fig. 6.2c. Also housed in the triangular arrangement of insulated sections at the end of the stalk is a closing resistor to prevent over-voltages in excess of the insulation level when connecting to EHV and HV transmission systems (2, Fig. 6.2c). The horizontal member of the triangle houses the contactor for switching the resistor in and out of the circuit.

types	ELF	SF 2-1	2-1 SL 3-1 4 1	4-2 SL 5-2 6-2	7-4 SL 8-4
breaker arrangement					
rated voltage kV insulation level kV rated current A breaking current kA breaking time cycles		123–145 ≦650 3150 20–25 3.5	145 245 ≦1050 4000 31.5–40 3–3.5	245 420 ≦1425 4000 40–50 2/3	550–800 ≦2100 2000...4000 40–50 2/3

a

Fig. 6.2 *Circuit-breaker Assemblies*
  a Live-tank assemblies (Fig. 10, Dunokel and Voigt, 1985. Copyright Brown Boveri)

Impact of SF_6 technology upon transmission switchgear 83

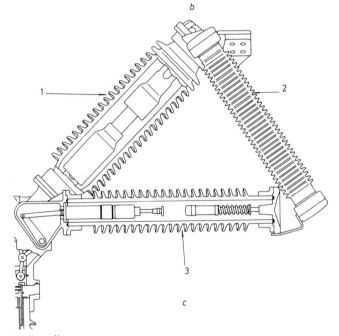

	ELK0	ELK1	ELK2	ELK3	ELK4
typical bay arrangement					
rated voltage (kV)	72-145	145-245	245-362	362-550	550-800
insulation level (kV)	650	750-1050	1050-1300	1425-1550	1800-2000
breaking current (kA)	40	40/50	40/63	40/63	40/50
rated current (A)	2000/3150	3150/4000	3150/4000	4000/6300	5000/6300

Fig. 6.2 *(continued)*
b Dead-tank assemblies–enclosed terminations (Fig. 1, Morii et al., 1978. Copyright CIGRE)
c Assembly of circuit breakers fitted with closing resistor (Fig. 9, Leupp, 1983. Copyright Brown Boveri)

An entire range of circuit breakers such as those shown on Fig. 6.2a is usually based upon a modular system, which permits all circuit breakers to be assembled from a small number of components.

Dead-tank circuit breakers tend to be mechanically simpler than live-tank breakers since the linkage between interrupters and driving unit is inherently shorter. However, in the dead-tank construction the dielectric stress to the earthed tank, as well as across the contact gap during recovery, must be considered seriously at the design stage. Short-circuit tests of a single interrupter unit will not fully stress a multi-break construction to earth, so that some proving tests must be undertaken to check the fully stressed condition at maximum interrupting currents.

Dead tank circuit breakers are manufactured by several countries worldwide covering voltage ranges from 36 kV to 550 kV.

An attraction of the dead-tank circuit-breaker concept is that it may be conveniently used as an integral component of completely metalclad installations and housed indoors. With such installations all the components of the substantion system (busbars etc) are enclosed and insulated with SF_6 gas. A range of such indoor dead-tank circuit breakers covering voltage levels from 72 to 800 kV, and which is typical of such breakers produced worldwide, is shown in Fig. 6.2b. The circuit breaker is the main vertical member of the system. At the lower voltages (72–245 kV) the transmission connection is by insulated cable, whilst at the higher voltages (245–800 kV) The connection is via a gas-filled bushing. Two sets of gas-insulated busbars are connected to the other side of the circuit breaker. Whereas at voltages above about 145 kV the busbars and interrupters of each of the three phases are housed separately (phase segregation), for voltages up to 145 kV the busbars and interrupters of all three phases may be housed within a single enclosure. (Although most manufacturers utilise three phases in one enclosure principally on account of reduced costs and space requirements, Sprecher & Schuh manufacture circuit breakers with a single phase per enclosure to limit damage in the event of an internal flashover and to reduce dynamic forces on the conductors during a short circuit).

In the dead-tank interrupter units shown in Fig. 6.2b the interrupting-chamber column is as symmetrical as possible, resulting in the mechanical forces being evenly distributed throughout the support structure to minimise the loading of the moving parts and to optimise transmission of the operating forces. A grading capacitor for the even distribution of voltage across the two breaks is also mounted within the circuit-breaker housing. Closing resistors are normally mounted in separate SF_6-filled chambers.

6.3 Metal clad installations

The world's first EHV metalclad substation at 132 kV was installed in Scotland by NEI (Reyrolle) in 1932, but was oil rather than SF_6 insulated. This was in successful operation for about 40 years, but despite being a technical 'first', it was not economic compared with open-type substations for normal applications (Ali and Headley, 1984). A limited number of metalclad installations relying on compressed air or Freon as the insulating medium followed during the next three decades, but it was not until the late 1960s, when SF_6 came to be fully accepted, that gas-insulated systems began to compete successfully with open-terminal layouts. The growth in the number of SF_6 metalclad bays installed by Mitsubishi of Japan for the period 1965–81 is shown in Fig. 6.3a. By the end of 1981, 1939 such bays had been commissioned by Mitsubishi.

The first 300 kV installation in the UK was a duplicate-busbar back-to-back arrangement which has given 10 years trouble free service (Ali and Headley, 1984). Further installations at 300 and 420 kV have followed. Many different designs of gas-insulated substations are now available throughout the world at voltages from 66 to 765 kV, with several trial installations at higher voltages.

The commercial significance of the gas-insulated-system concept is that, with SF_6 insulation, distances are reduced so that much more compact substation layouts become possible. These are easier and more economic to accommodate than open-type layouts on expensive land in urban areas, where power-consumption demands are large. Since the gas-insulated system needs to be interconnected at some point to a conventional transmission system, there are a number of options concerning the extent of the gas-insulated units. Whereas Fig. 6.3b(i) shows schematically the layout of an EHV conventional system, Fig. 6.3b(ii) shows a combination of conventional and gas-insulated components (hybrid GIS) and Fig. 6.3b(iii) shows a complete gas-insulated system. All three systems are fed from conventional overhead lines. Typical indoor gas-insulated systems are shown in Fig. 6.3b(iv). These Figures show that the ground area required for a hybrid GIS is approximately 25% of that for a conventional layout, whereas a complete GIS only requires about 7% of the conventional system space. For 145-kV systems with the three phases contained within the same enclosure, the space requirements can be as low as 5% of a conventional system (Merlin–Gerin). Of course, the economic gain of reducing the area of expensive land required is partly offset by the additional expense of providing

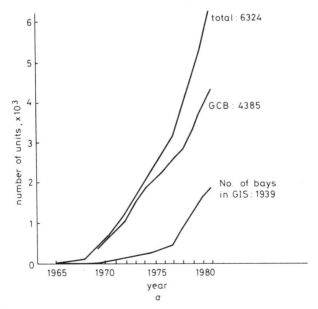

Fig. 6.3 *Metal clad installations*
a Growth in the number of Mitsubishi gas-insulated bays in Japan (Fig. 1, Kuwahara et al., 1983. Copyright IEEE)

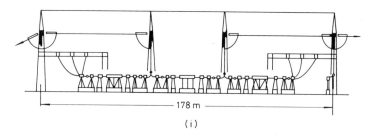

(i)

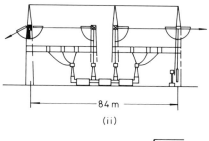

(ii)

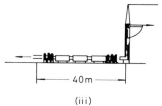

(iii)

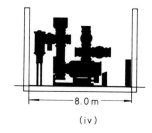

(iv)

b

Fig. 6.3 *(continued)*
b Comparison of size and appearance of conventional and gas-insulated systems (Fig. 3, Morii et al., 1978. Copyright CIGRE (Figs. 19, 18, GEC Brochure – SF_6 for Transmission and Distribution Publication 1294–74)
(i) Conventional switchgear
(ii) Hybrid GIS
(iii) GIS
(iv) 420 kV metal clad circuit breaker

Impact of SF$_6$ technology upon transmission switchgear

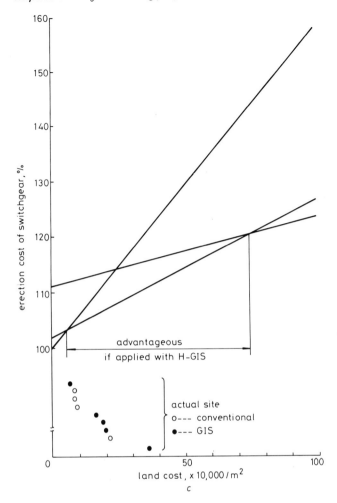

Fig. 6.3 *(continued)*
c Comparison of erection costs (Fig. 5, Morii et al., 1978. Copyright CIGRE)

total metallic encapsulation. A comparison of the erection cost as a function of land cost for conventional, hybrid and full gas insulation as applied to Japan (Morii *et al.*, 1978), in relation to 550 kV installations, is given in Fig. 6.3c. the hybrid GIS system clearly has advantages for low- and medium-cost land sites, whereas the complete GIS system has adantages for expensive sites. In addition, the advent of gas-insulated systems has made available sites such as underground caverns, which could not so easily be used economically or operationally with conventional systems.

SF$_6$-insulated metalclad systems have also proved to be more reliable and to require less maintenance than conventional open layout systems (Andersen and Simms, 1983). Mitsubishi claim that, with their 1939 commissioned gas-insulated bays, the rate of troubles per unit time has decreased to about 5% over a period of about 17 years, and that only five serious failures have been experienced in 30 000 unit years (none of which were due to the dielectric failure of the insultion) (Kuwahara et al., 1980).

6.3.1 Components and structure
The layout of a high-voltage substation involves a proper arrangement of the switchgear installation to provide a high degree of continuous and reliable service with the minimum of financial outlay. A number of basic arrangements are commonly employed nowadays, and these include single, double and multiple busbar shemes, as well as breaker and a half, 2-breaker and ring-bus schemes (Fig. 6.4a) (e.g. Stepinski, 1978). Brown Boveri have recently proposed two new configurations called 'crossed-ring' and 'ring-bus with bridging breaker' arrangements (e.g. Dunckel and Vogt, 1985), which would have superior flexibility compared with the more conventional systems. Reliability analyses indicate that these new proposed systems would give improved system reliability compared with the ring-bus scheme or the 2- and 1-breaker arrangements, which, in turn, are more reliable than single-, or double-busbar arrangements (e.g. Schmit and Klink, 1978; Kulik and Schramm, 1980, Koeppi et al., 1983; Dunckel and Vogt, 1985). Cost comparisons show that the ring-bus system is somewhat cheaper, but that the breaker-and-a-half system is marginally more expensive than the newly proposed systems.

The implication of these observations is that, since customer demand may be for one of a number of possible system configurations, the design of components of a metalclad system and their interconnections must be flexible in order to permit these various configurations to be composed economically. The complexity of such requirements with regard to the number and dispostion of circuit breakers, isolators and busbars, required to form the various bays and their interconnections, is illustrated by a comparison of the duplicate busbar system, the ring-bus scheme and the breaker-and-a half arrangement shown schematically in Fig. 6.4a.

A section of a single bay of a typical metalclad system with double busbars and cable termination is shown in Fig. 6.4b. (This particular system is rated at 145 kV/40 kA (3·15 kA) with an insulation level of 650 kV peak.) Typical components of such a system include a dead-tank circuit breaker, a number of disconnectors, several earthing switches, busbar units, current and voltage transformers and cable terminations. In the system shown, the circuit breaker is of the dead-tank variety with all three phases housed within the single enclosure. Each interrupter is surrounded by an insulating enclosure which prevents the propagation of ionised gases from one pole to another. The use of a vertical rather than horizontal tank clearly provides advantages for interconnection with the remainder of the system. Furthermore, the vertical-tank construction has the important feature that any particulate contamination produced by arcing, and which can cause premature electrical breakdown, will

Impact of SF_6 technology upon transmission switchgear

fall to the bottom of the tank to a region of zero dielectric stress (Ali et al., 1982; Ryan and Watson, 1978).

Housing the busbars of the three phases of the power supply within a single enclosure is economically attractive but introduces major technical problems associated with phase-to-phase insulation, interaction between phases during high-current interruption and more severe thermal and mechanical stresses. The most severe phase-to-phase-insulation demand is likely to occur when a lightning surge appears on one phase when the potential of another phase might be at the maximum of the opposite AC polarity (Azumi et al., 1980). This requires that the phase-to-phase insulation level should be at least 1·2 times the lightning impulse-withstand level of the system. In order to utilise the space in the SF_6 enclosure most effectively for high-voltage insulation, the busbars of the three phases are arranged in a right-angle isosceles-triangle configuration.

An example of the complex electric-field distribution within such a 3-phase enclosure is shown in Fig. 6.4c. This corresponds to 3-phase rated voltage on the three conductors and an impulse equivalent to the lighting-impulse-withstand voltage applied to conductor 3, with the enclosure maintained at earth potential. To reduce the number of spacers in a system, the busbars may be supported by insulators in the wall of the enclosure.

The process of disconnecting parts of a substation system involves firstly interrupting a normal load current before operating the disconnector and then the earthing switch. Load-current interruption in the past was achieved by opening the circuit breaker. The disconnector is then required to interrupt the small capacitive charging currents ($\sim$ 300 A in Japan) that may flow through the short lines as far as the open circuit breaker. In Europe and North America such disconnectors are slow operating with an arc duration of 0·33 s, whereas in Japan faster operation is common (e.g. Boggs and Fujimoto, 1985). The length of the disconnector gap is determined by insulation-co-ordination requirements (Section 6.3.2), so that the gap will not flash over to the earthed enclosure.

Switching duties during disconnection procedures are not well specified (e.g. Spindle, 1980), but are more onerous in an SF_6 metalclad system than in a conventional air insulated system (Cuk et al., 1980). As system voltages continue to increase worldwide, the demands made of such switches are also likely to increase (Yanabu et al., 1985). The switching duties during load breaking are different from fault interruption, so that economically it may be advantageous to plan systems with certain interrupters designed specifically for load breaking. For instance, on the Dinorwig pumped-storage scheme six switch disconnectors and four circuit breakers are used instead of nine circuit breakers. At St. John's Wood, London, switch disconnectors have been installed instead of disconnectors to reduce the amount of plant shut down during various operations (Harris and Simms, 1978).

An indication of the duties demanded of such switch disconnectors may be obtained by considering the duplicate busbar system shown in Fig 6.4a(i). This involves four switch disconnectors incorporated in a closed loop. If any one of

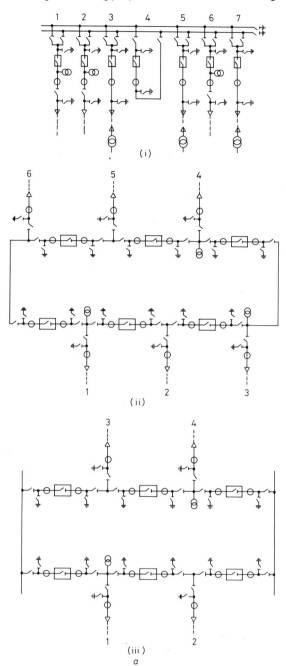

Fig. 6.4 *Structure and components*
 a Basic system arrangements (Fig. 2, Schmitz and Klink, 1978)
 (i) Double busbar
 (ii) Ring-bus.
 (iii) $1\frac{1}{2}$ circuit-breaker system

Impact of SF_6 technology upon transmission switchgear 91

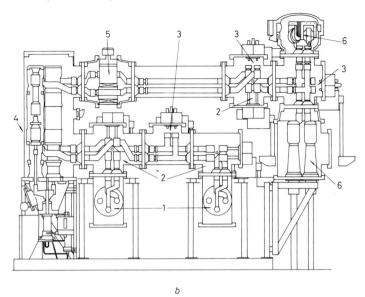

b

Fig. 6.4 *(continued)*
b Section of a typical bay (p. 3, Merlin–Gerin E/A462)

the disconnecting switches is opened it will be required to interrupt part of the closed-loop current. The switching duty is therefore typically 50% of the rated continuous current of the bay. In Japan the maximum value of this current is 8 kA in a 300 kV system and 12 kA in a 550 kV system. The recovery voltage V_R across the disconnecting switch is (Morita et al., 1985)

$$V_R = KIl \times 10^{-3} \text{ volts} \tag{6.1}$$

where I is the current in amperes and l is the length of loop in metres. K is a constant determined by the busbar material and composition, being typically of the order of $0 \cdot 1 - 0 \cdot 5 \, \Omega/\text{km}$. As a result, the recovery voltage lies in the range from several tens to several hundreds of volts, depending upon the layout of the gas-insulated system. In Japan the maximum recovery voltage is specified as 300 V (Morita et al., 1985; Suzuki et al., 1984). The disconnecting switch duty is therefore characterised by a relatively low value of recovery voltage and a high interrupting current.

With modern substation systems, earthing switches may also be required to switch currents under some operating conditions. These correspond to a current being induced electromagnetically in a line earthed at both ends from a neighbouring energised line (Suzuki et al., 1984). Typically for a 550 kV 12 kA transmission system, an earthing switch may be required to interrupt 1·5 kA at 90kV, whilst for 300 kV 8 kA lines, the duty is 1 kA at 50 kV (Suzuki et al., 1984).

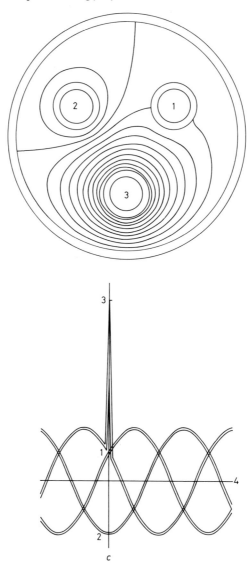

Fig. 6.4 *(continued)*
c Electric-field distribution in a 3-phase enclosure and voltages on each phase with lightning surge on phase 3 (Fig. 7, Sprecher and Schuh Energy brochure)

Thus an earthing switch is required to interrupt a relatively low current with a high recovery voltage.

Since the life expectancy of a gas-insulated system is prolonged, both disconnecting and earthing switches need to have the capability of switching many

Impact of SF_6 technology upon transmission switchgear 93

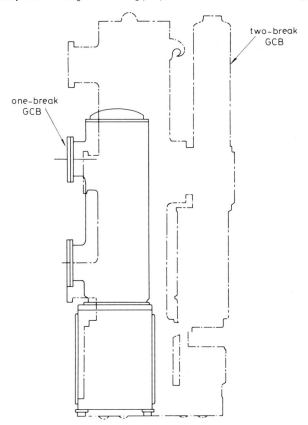

Fig. 6.4 *(continued)*
d Reduction in size by adopting a one break circuitbreaker (Fig. 9, Ikeda et al., 1984. Copyright IEEE)

times ($\sim$ 100–200 times in Japan) (Suzuki *et al.*, 1984) without maintenance of contacts and without a resulting accumulative deterioration of insulation strength. Both these factors may be advantageously affected by limiting arcing duration during the interruption process, whilst simultaneously meeting the switching-performance requirements. Ideally, disconnecting-switch and earthing-switch designs should be separately optimised to meet their different switching duties. Harris and Simms (1978) describe a monoflow puffer interrupter used as a disconnecting unit on the Dinorwig pumped-storage scheme and at St. John's Wood, London. More recently Suzuki *et al.*, (1984) have investigated the possibility of using newer and simpler forms of SF_6 interrupters such as rotary-arc and self-pressurising units as disconnecting and earthing switches. A comparison of the suitability of these types of interrupters for disconnecting and earthing duties is more appropriately considered in conjunction with their use for distribution-system applications (Chapter 9).

Voltage and current transformers are mounted within the metal clad system, and are therefore SF_6 insulated. The voltage transfromers are normally 3-phase electromagnetic in nature, whilst the current transformers are normally of the 3-phase type with one or more cores. SF_6 by-products are normally removed by absorbants placed in the circuit-breaker compartment. Typically all compartments are pressurised to 3·5 bar (gauge) at 20°C, except the circuit-breaker compartment, which is at about 6·5 bar.

Aluminium is generally preferred to steel for the manufacture of the various enclosures. Steel suffers from substantial electromagnetic losses, so that the enclosure diameter is often determined by these losses rather than the dielectric strength of SF_6 (Hoegg, 1980). Furthermore, aluminium can be easily cast into optimal dielectric shapes and requires no special corrosion protection. Cast rather fabricated enclosures have produced cost reductions [Harris (GEC), private communication]. The lower mass density of aluminium also leads to low-weight enclosures.

The metal clad system is divided into a series of sealed compartments to ensure internal separation of functions. This limits any internal faults or SF_6 leaks to a single compartment, and facilitates pressure monitoring and leak and fault detection. Optimisation of the number of enclosures containing different numbers of apparatus can lead to a reduction in cost [Harris (GEC), private communication]. Compatible with reduced frequency of maintenance (typically greater than 10 years) sufficient dielectric strength within the units of a metal clad system is usually assured even if the SF_6 pressure decreases to half the rated level (e.g. Eidinger and Petitpierre, 1979). Such loss of pressure is very improbable with currently available sealing systems. Loss of dielectric strength due to particulate impurities can be avoided through the use of particle traps (e.g. Spindle, 1980). However, there is evidence that such measures are unnecessary if the system is properly designed.

Misalignment of the various sections of a metal clad system may occur owing to a build up of manufacturing and site tolerances, differential thermal expansion resulting in the movement of sections relative to each other, and the possible movement of foundations across the switchgear site (Ali and Headley, 1984). Such misalignments may be tolerated through the use of bellows. Standard items for connecting a metal-enclosed system to overhead lines, cables, transformers or reactors are also required. Connections to overhead lines are made through air/SF_6 bushings. The bushing is filled with SF_6 and the enclosure is fitted with refilling and safety accessories.

Because of the permutation of the manner in which the various metalclad components may need to be combined in order to cater for the variety of substation layouts demanded by customers, manufacturers have increasingly utilised the unit-section component concept. This has enabled manufacturers to

Impact of SF_6 technology upon transmission switchgear

limit the number of components, with a consequent improvement in quality control and a reduction in overall cost.

The evolution of the SF_6 gas-insulated system has relied, and is likely to continue to rely, upon two major factors. The first is associated with the concept of an integrated system, of which the integration of a separate phase system into a single 3-phase enclosure is an example. The second involves minimising the size of each component of the system, of which the decrease in size of the circuit breaker, consequent upon the reduction of the number of breaks per phase, is a major consideration. The reduction in size of the circuit breaker, which results from the use of a single- rather than a double-break interrupter, can also lead to a reduction in the size of the interconnected metal clad system. Indeed, the non-availability of a single-break-per-phase circuit breaker at the higher transmission voltages has hitherto been a major factor in limiting the use of the three phases in a single enclosure to the lower transmission voltages (e.g. Azumi *et al.*, 1980). The difficulties of testing to prove the rating (Chapter 9) and the consequences of an interphase fault are also major reasons for limiting three phases in one enclosure to the lower voltages.

6.3.2 Insulation co-ordination

The basic aim of insulation co-ordination is to dimension the insulation of a system in accordance with operational conditions to ensure safety and reliability economically. The co-ordination process is based upon a consideration of overvoltage stresses, the electrical strength of components, overvoltage protection and the desired degree of safety against overvoltages. Since IEC recommendations allow a wide latitude for obtaining an economically optimum layout, there is no predetermined method for designing a network. The optimum technical and economic solution must be found iteratively by examining all possible alternatives (Burger, 1979). Although insulation co-ordination is already well established for conventional open-layout substations, it is still incomplete for SF_6-insulated systems on account of their basically different nature. The characteristics of SF_6-insulated systems, which differ from those of conventional systems, include comparatively low surge impedance and propagation velocity, small spread of buses and equipment, reduced ability for self-restoration following insulation breakdown, and different overvoltage/time characteristics.

Overvoltages which exceed the insulation level of the system can lead to flashovers. Consistent with the principles of insulation co-ordination, it is not an economic proposition to raise the insulation level of power systems to such an extent as to withstand all possible overvoltages. Instead the overvoltages must be restricted to a certain level (e.g. Ruoss, 1979), either by limiting the overvoltages at source, by using surge arresters or through a combination of both.

Three major types of overvoltages may be identified on power systems (e.g.

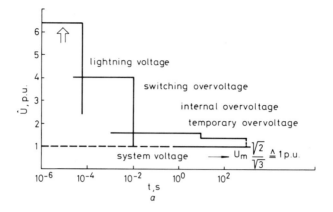

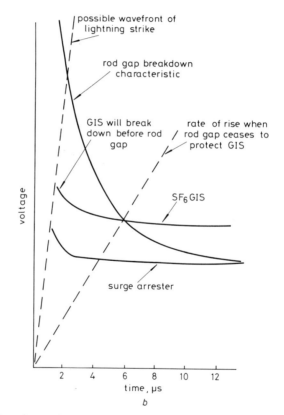

Fig. 6.5 *Insulation of co-ordination*
a Duration of various overvoltage stresses (Fig. 2, Burger, 1979. Copyright Brown Boveri)
b Voltage: time characteristics for rod gap, gas-insulated system and surge arrester (Fig. 2, Andersen and Simms, 1983)

Impact of SF_6 technology upon transmission switchgear 97

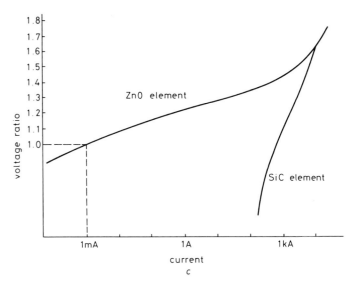

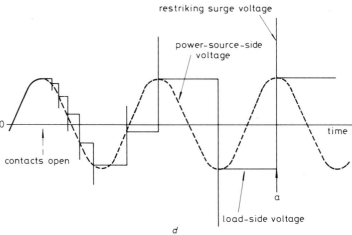

Fig. 6.5 *(continued)*
c Volt: ampere characteristics of ZnO and SiC arresters (Fig. 7, Oshima et al., 1983)
d Explanatory diagram of disconnecting switch restriking surge (Fig. 1, Ogawa et al., 1985. Copyright IEEE)

Ruoss, 1979; Mayer, 1983). These are atmospheric overvoltages (caused by lightning strokes), transient overvoltages (due to switching operations) and temporary overvoltages (due to network-resonance phenomena etc). Each of the overvoltages may be characterised by its rise time, maximum voltage level and associated energy. Fig. 6.5a (Burger, 1979) shows the voltage levels and duration of the three major types of overvoltages. Of these, the atmospheric overvoltages constitute a particularly severe problem, with their associated high voltage level and rapid waveform rise time, which may be less than 1 µs. At the other extreme, the temporary overvoltages, despite their relatively low voltage levels, are associated with very high energies on account of their prolonged duration. Whereas system-induced overvoltages may be limited at source by taking suitable precautions, it is impossible to prevent lightning overvoltages. The risk to plant due to the latter depends upon the frequency of storms and the proximity of a strike to a transmission line to the plant to be protected. Overvoltages produced owing to premature arc extinction leading to current chopping can occur during the switching of low inductive currents and are considered further in Section 7.1.

In the UK overvoltage protection of conventional air-insulated substations has almost always been based upon the use of rod gaps in air (Simms and Andersen, 1983). Although such protection is imprecise, it is inexpensive, and, when used in conjunction with delayed auto-reclosing of the circuit breaker allows the supply to be self-restoring. The flashover-voltage/time characteristic for a rod gap in air is of a similar shape to that for the long air gap across the porcelain insulation of a system which it is protecting (Fig. 6.5b). However, SF_6-insulated metalclad substations cannot be protected by rod gaps for two basic reasons. First, a flashover of a co-ordinating gap situated within the gas-filled cladding is not acceptable, since the presence of the SF_6 co-ordinating gaps could impare the dielectric integrity of the GIS and the chamber would become polluted and possibly damaged. Secondly, the flashover-voltage/time characteristic for concentric structures, such as those used in SF_6-filled systems, is unlike that of long air gaps. For times greater than about 3 µs, the flashover voltage for the SF_6-insulated equipment is nearly constant with time (Fig. 6.5b) and is less than that of the rod gap at short times. Consequently a rod gap in air, placed across the air-insulated entrance bushing to an SF_6 metalclad substation, will not give protection against lightning surges with rapid rates of rise.

The alternative method of protection is with co-ordinated surge arresters, when the protective characteristic of the arrester can be shaped to be always below the gas-insulated-enclosure withstand-voltage curve (Andersen and Simms, 1983). Recent developments in metal-oxide techinology have made metal-oxide resistors attractive for such power-system protection. A typical current/voltage characteristic for such a resistor is shown on Fig. 6.5c. The normal operating point is arranged to be at a voltage ratio of unity on the characteristic. As soon as the voltage across the surge arrester increases, the operating point moves along the charactertistic so that an increased current flows without delay to reduce the overvoltage. In solidly earthed systems the operating point can be set accurately below the unity-ratio value, so that lightning and switching overvoltages can be limited to 5–10% and 25–30%, respectively, of the levels achieved with more conventional arresters. The characteristics shown in Fig. 6.5c indicate the improved performance of ZnO compared with SiC in limiting the unity voltage-ratio current to 1 mA as opposed to 1 kA. This had led to the series gap, used with SiC to limit the normal current flow from 1 kA, being made redundant by the introduction of ZnO resistors. Furthermore, a metal-oxide resistor is able to absorb about 40% more energy than a silicon-carbide resistor of the same volume. The metal-oxide element also has a sufficiently long life to be compatible with long-duration-periods between servicing.

With SF_6-insulated systems, the metal-oxide resistors can be mounted directly inside an SF_6-filled vessel. If the length of the SF_6 trunking between the overhead-line termination and the power transformer is greater than 20–30 m, additional surge arresters may need to be installed at the overhead-line/gas-insulated-system junction and also adjacent to a power transformer (Simms, 1984). Using such considerations, the Drakensberg 420 kV SF_6-insulated substation, which incorporates a 150 m vertical SF_6-filled busbar system, is claimed to be properly protected even against direct lightning strokes occurring within 1 km of the substation.

Switching overvoltages which occur when energising or de-energising transmission lines have increased in significance with the introduction of transmission voltages of 525 kV and above. Such overvoltages, which can be produced during circuit-breaker closure, may be limited by inserting a current-limiting resistor during the switch-closing operation, as described in section 6.3. With SF_6 metal clad systems this resistor is mounted within an SF_6 enclosure to improve insulation co-ordination of the system.

There is documented evidence that the operation of disconnecting switches in metal clad systems can also produce overvoltages leading to flashovers. For instance, of four flashovers recorded on the 500 kV Mica gas-insulated installation of British Columbia Hydro between 1976 and 1980, one occurred in the disconnect chamber following substantial arcing between the switch contacts (Cuk et al., 1980). Nishiwaki et al., (1983) have observed a similar fault to ground under test-station conditions.

Restriking surges across the contacts of a disconnect switch occur a number of times before interruption is complete. The slow-operating European and North American disconnectors produce a relatively large number of small transients, whilst a fast disconnector of the type used in Japan produces fewer but larger transients with clearance occurring at randomly different contact separations. The voltage transients assume the form of a step-like voltage waveform on the load side (Fig. 6.5d; Ogawa et al., 1985) with restriking surges of frequency in the range of several hundred kilohertz to several megahertz. The flashovers to ground, observed during operation of disconnect switches, have been shown to be due to these high-frequency surge voltages (Ogawa et al., 1985). The significant feature of these flashovers is that the flashover voltage is appreciably reduced below the normal withstand voltage to ground owing to the persistent influence of arc-heated SF_6. It is to overcome these problems that the use of more efficient disconnecting switches has been proposed by Yanabu et al., (1982) (Section 7.8).

A knowledge of the disconnecting-switch-induced surges would appear to be required for proper insulation-co-cordination studies of gas-insulated substations. For this reason, computer simulation of such surges has been reported by Ogawa *et al.*, (1985) for 111 different circuit configurations with 10 different gas-insulated substations.

6.3.3 Internal arcing faults in metalclad enclosures
Faults in metalclad systems can lead to arcing within the restricted volumes of the enclosures, which in turn may cause rupture (burnthrough) of the enclosure wall. Not only can this reduce the reliability of gas-insulated systems, but it may also produce safety hazards for operating personnel. Although such occurrences are expected to be rare, because of the safety implications and prolonged out-of-service repair period, manufacturers and utilities worldwide have been concerned about the time interval that these enclosures are able to withstand an internal arc.

A survey of faults on gas-insulated systems in North America in 1979 showed that, of 44 in-service faults, three involved burn-throughs with an average fault current level of 13 kA (Chu *et al.*, 1982). During an 18-month period up to 1982, Ontario Hydro experienced nine in-service faults in its four SF_6 substations, involving a maximum fault-current level of 31 kA. There is evidence that fault performance of installations outside North America is substantially better (Chu and Tahiliani, 1980). Siemens claimed in 1982 that, of their 2000 gas-insulated bays (each with 6–7 circuit breakers) in service worldwide for more than 12 years, there was an average fault rate of a few tenths of a per cent per breaker year, which is substantially less than the North American experience (Diessner and Schramm, 1982). A joint investigtion by ASEA and the Swedish State Power Board of conventional substations yielded a total failure rate of 0·025 per substation (Ericsson, 1982), whilst Japanese statistics indicate a total failure rate of 0·006 per gas-insulated-system year for the period 1973–78 Wakanishiv *et al.*, 1982).

Rupture of the metal clad enclosure may occur due to either over-pressurisation caused by arc heating or burn through due to the action of the arc root on the enclosure wall. Fig. 6.6*a* (Chu *et al.*, 1982) shows arc formation between a live busbar and the enclosure wall following a fault. In general, the arc will be driven axially and azimuthally by the electromagnetic forces produced by the various current flows (cf Section 4.3). There are, however, two situations when the arc may remain stationary; the first corresponds to the arc having been driven electromagnetically against an insulating spacer and the second to the electromagnetic forces being balanced due to equal and opposite current flows (Fig. 6.6*a*). The burn-through time t_B for this situation is the sum of the time during which the arc is mobile, t_W, and the burn-through time once the arc becomes stationary, t_{BS}; i.e.

$$t_B = t_W + t_{BS} \qquad (6.2)$$

Experiments indicate that the burn-through time for a stationary arc increases with the thickness of the enclosure wall D and decreases with the arc current I (Stasser et al., 1975); i.e.

$$t_{BS} = CD/I \ milliseconds \tag{6.3}$$

where C is a constant of proportionality which depends upon the thermal diffusivity of the wall material (Diessner and Schramm, 1982) and the equipment size ($C = 540\,\text{ms}\,\text{kA}\,\text{mm}^{-1}$ for a 420 kV-size aluminium coaxial chamber (Pettersson and Graustrom, 1977)). Some researchers find a stronger dependence of t_{BS} upon D than that given by eqn (6.3) (Diessner and Schramm, 1982), whilst others attempt to relate t_{BS} to more fundamental properties, such as the metal volume heated by the arc and the amount of energy injected into the metal (Kuwahara et al., 1982).

For an axially mobile arc, t_W depends upon the separation L of the spacers in the enclosure and the axial velocity v, which experiments indicate is linearly proportional to the arc current (Chu et al., 1982); i.e.

$$t_W = L/v = L/(\alpha I) \tag{6.4}$$

where α depends upon the size of the equipment and the gas pressure. The burn-through time t_B decreases with gas pressure and may be increased by a factor of $2 - 2\cdot 7$ owing to azimuthal arc motion (Diessner and Schramm, 1982). The burn-through times of aluminium enclosures can be of the same order as steel if properly designed (Hoegg, 1980).

In order to estimate the likelihood of a burn-through in a given substation system, it is essential to know the fault statistics of that system, in terms of the probability density of a fault current occurring, and the probability density for the number of current-waveform cycles required to clear a fault by the circuit breaker. A typical fault-current distribution projected for Ontario Hydro's 230 kV Clairville substation indicates a most probable fault current of about 47 kA whilst a typical distribution of fault clearing times indicates a peak between 3 and 5 cycles (Chu et al., 1982). Burn-through will, of course, only occur if the burn-through time (eqn. 6.2) is less than the fault-clearing time. Typical results for the burn-through probability per fault as a function of the peak of the fault-current distribution is shown for an aluminium enclosure in Fig. 6.6b. Clearly, if future systems are to cater for higher fault-current levels the probability of burn-through is likely to increase unless design changes are made.

The pressure rise due to arc heating which may cause the pressure relief disc to operate is determined by considerations similar to those given in Sections 4.1 and 4.2. Empirically this pressure rise is given by (Chu et al., 1982)

$$\Delta p = \beta \frac{Ut}{V} \ kilopascals \tag{6.5}$$

where U is the arc voltage, t the arc duration and V the enclosure volume. β is a constant of proportionality empirically determined as 60 kPa/kJ (e.g. Chu et al., 1980). This expression is valid up to 20 kJ/litre of arc energy (Chu and Law,

1980) and is consistent with the assumption that, if there is a uniform distribution of thermal energy throughout the gas-filled enclosure, then 70% of the arc energy is absorbed by the gas. These estimates are for SF_6 in aluminium enclosures, and the 70% value includes the contribution from the exothermic reaction which occurs between SF_6 and aluminium. There is evidence that,

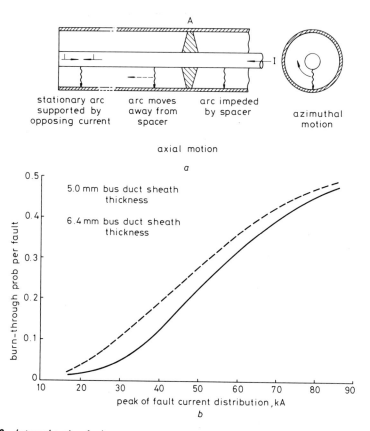

Fig. 6.6 *Internal arcing faults*
  a Arc behaviour during a fault (Fig. 1, Chu et al., 1982)
  b Burnthrough probability for various fault-current distributions (Fig. 7, Chu et al., 1982. Copyright IEEE)

owing to this exothermic reaction, the pressure rise with aluminium enclosures of small dimensions may be greater than for other materials (Kuwahara *et al.*, 1982). However, for larger enclosure dimensions (> 8 cm arc length), the ratio of chemical reaction to arc energy becomes small and the difference between aluminium and other materials becomes negligible.

Some sample predictions for the probability of the pressure-relief disc rupturing per fault as a function of over-pressure have been made by Chu et al, 1982. These predictions neglect arc-induced pressure waves (e.g. Leclerc *et al.*, 1980), which have been shown to be important in promoting local pressure differences in a similar environment (e.g. Lutz and Pietsch, 1982).

The results of such calculations, based upon the North American experience, suggest that there is a 25% probability that burn-through may occur in the event of a fault, such a probability being greater than that of rupturing a relief disc ($\sim 15\%$). The probability of a burn-through occurring on the North American system is, on average, one ocurrence every 13 years, whilst the chance of personnel being exposed to the by-products of a burn-through is 0·015 per annum. Although there are differences in some of the statistical data for North America compared with the rest of the world, as indicated above, the final conclusions concerning the danger to personnel are remarkably similar in predicting a tolerably small risk (e.g. Ericsson, 1982) and relatively rare burn-through occurrences.

6.3.4 Internal maintenance requirements and reliability

Since SF_6 circuit breakers and SF_6-insulated systems have been in operation for over 15 years, information from actual inspection in the field concerning internal maintenance requirements is now becoming available. Of the various factors which are expected to determine maintenance requirements, it is contact wear due to repeated current interruptions which is the dominant factor (Kuwahara *et al.*, 1983*a*.) PTFE nozzle wear, degradation of SF_6 purity, condition of the impurity absorbants, wear of O-ring seals and degradation of solid insulation by arcing by-products are apparently not a problem. These investigations suggest that the periodic internal inspection of SF_6-filled circuit breakers may be extended to once in 12 years or more.

Methods for analysing the quality of the SF_6 gas within a metalclad system are discussed in Section 9.1.3. Suzuki *et al.*, (1982) have investigated the degradation of sealing grease due to the arcing by-products of SF_6.

The accumulated failure rate per unit year for SF_6-insulated systems in Japan has decreased from about 0·0175 since their introduction in 1970 to about 0·006 for the period 1973–78 (Nakanishi, 1982). This improved reliability has been achieved as a result of a reduction in the number of mechanical faults through improved factory-testing procedures, improved gas-sealing technology, standardisation of parts and components, and the adoption of installation and maintenance criteria derived from in-service experience (Nakanishi *et al.*, 1982).

6.4 Artificial current zeros

In alternating-current systems, circuit interruption is sought when the current passes through zero at the end of the current half-cycle in order to limit system overvoltages and circuit-breaker demands. However, there are situations where a natural current zero is either delayed or does not occur. These correspond to high degrees of asymmetry of the sinusoidal current waveform due to the superposition of a prolonged 'DC' current decay or due to direct-current flow. The first case corresponds to short-circuit fault currents from high-power generators under some adverse conditions. The second case corresponds to conditions of high-voltage direct-current transmission. In both cases circuit interruption is sought by inducing artifical current zeros.

6.4.1 Generator circuit breakers

The use of generator circuit breakers in power plants leads to considerable advantages and savings in station layout and operating procedures, thus increasing the overall power plant reliability.

The operation of a generator–transformer unit (Fig. 6.7a) in the past required a start-up transformer which obtained its supply from the high-voltage system or from a separate network. Once the generator reached the synchronous speed, it was connected to the network by means of the high-voltage circuit breaker. The station supply was immediately switched from the start-up transformer to the unit service transformer, which branched off between the main transformer and the generator. The incorporation of a generator breaker between the generator and the transformer (Fig. 6.7a) allows power at starting and shut down to be drawn directly through the main transformer and the unit service transformer. Hence there is no need for the change-over from the auxiliary supplies after the generator has been synchronised.

Fault protection of the generating plant is achieved either with the circuit breaker installed in the busbars between the generator and transformer or the circuit breaker on the high-voltage side of the transformer (Fig. 6.4a). Fault interruption is sought to be as early as possible, in order to limit possible

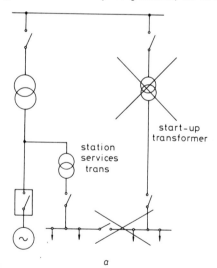

Fig. 6.7 Generator breakers
    a Generator circuit layout (Fig. 7, Dunckel and Voigt, 1985. Copyright Brown Boveri)

Impact of SF_6 technology upon transmission switchgear

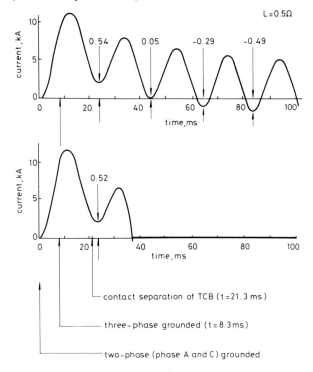

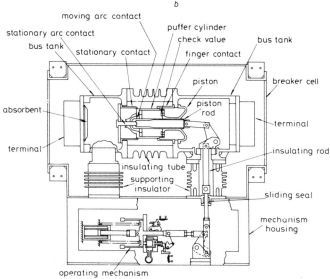

Fig. 6.7 *(continued)*
b Zero-miss current waveforms with and without interrupter arc (Fig. 2, Ishikawa et al., 1984. Copyright IEEE)
c Section of puffer-type generator circuit breaker (Fig. 3, Mitsubishi brochure HBN-84785)

generator damage. However, in practice a decaying DC component may, under some conditions, be superimposed upon the AC component. The decay of the DC component is determined by the ratio X/R (X = generator reactance, R = stator resistance) (Thuries et al., 1980), so that for large values of X/R the occurrence of a current zero may be substantially delayed (Fig. 6.7b; Ishikawa et al., 1984). For small generator plants, the value of X/R is sufficiently low to give a time constant of the DC component which is short ($\sim$ tens of milliseconds) compared with the breaker opening time. However, in larger generator plants $X/R \sim 80$–200, which can, in principle, lead to current-zero delays of several hundred milliseconds. Under such conditions, the time delay to current zero needs to be artificially shortened in order to prevent excessive plant damage under fault-current conditions.

The rating of large power plants is currently in the range 600–1650 MVA (Thuries et al., 1980). Power is usually generated at a voltage of 16–24 kV with currents in the range 7–50 kA, depending upon the type of power source (thermal, nuclear, pumped storage etc). Generator short-circuit currents are in the range 50–180 kA, whilst the short-circuit currents supplied on the output side of the transformer may be in the range 100–380 kA (Thuries et al., 1980). In the case of pumped-storage power sources, additional demands are made owing to the change-over from power production to raising water being required to alternate routinely; e.g. about twice daily (Morii et al., 1979).

These factors impose more onerous demands upon circuit breakers in certain respects than do transmission applications. For instance, because generator breakers are required to carry much higher rated currents, the circuit-breaker design must be such that contact resistance is low, the breaker needs to physically fit into the generator–transformer chain and provision needs to be made for forced convection cooling if neessary. The circuit-breaker requirement to interrupt much higher currents (but at limited voltages) than do conventional circuit breakers may be met through the use of larger-diameter nozzles to accommodate the larger-cross-section arc, or a higher gas pressure to limit the arc cross-section. The decay of the DC component under short-circuit conditions, which leads to a delayed current zero (e.g. Morri et al., 1979; Kulike and Schramm, 1980; Thuries et al., 1980) may be accelerated by increasing the generator-side resistance R naturally as a result of establishing the arc in the circuit breaker. The circuit-breaker design therefore needs to be optimised so that the arc resistance is rapidly maximised to produce the earliest possible current zero, and also so that the corresponding energy dissipated in the interrupter due to any prolonged arcing can be absorbed. The demands made by pumped-storage systems for frequent change-over requires the circuit breaker to interrupt rated currents repeatedly, so that the circuit-breaker life needs to be extensive and component wear minimised.

Although generator circuit breakers have traditionally been of the air-blast type (e.g. Thuries et al., 1980), the above demands are met more by suitable circuit-breaker design rather than by a choice between air and SF_6 as arcing media. However, there may be benefits to be gained through the faster recovery of SF_6, and there has recently been a trend to instal SF_6 breakers for generator protection. A 2-pressure (15/3 kg cm^{-3}) full-duoflow SF_6 circuit breaker was recently installed by Mitsubishi at the Okuyoshino pumped-storage station of the Kansai Electric Power Co. (Morii et al., 1979). By keeping the ohmic resistance across the contacts to 3·5 $\mu\Omega$, the breaker is able to carry currents up to 16 kA without any forced cooling. Following contact separation, an additional reistance of, typically, 10 mΩ appears in the circuit due to arcing, the value being greater at lower peak currents (Morii et al., 1979). In order to insert this resistance as rapidly as possible (so that the current-zero delay is minimised to produce efficient interruption and reduce energy dissipation in the interrupter), it is advantageous to open the circuit-breaker contacts quickly using efficient operating mechanisms. A reduction of the delay to current zero from 62 to 36 ms has been achieved (Fig. 6.7b; Ishikawa et al., 1984) with minimum and maximum arcing times of, typically, 0·2–0·7 cycles (Morii et al., 1979), which limits energy dissipation in the interrupter to about 20–250 kW s, respectively (Morii et al., 1979). Contact-erosion measurements on a single-phase unit (with allowance for interruption of the second and third phase 0·25 cycles later) suggest that the contacts would withstand about 2700 load-switching repetitions (Morii et al., 1979), and so meet the estimated demand for an inter-maintenance period of 3 years for a pumped-storage system requiring power interruption twice daily (Morii et al., 1979). More recently Mitsubishi have introduced a single-pressure puffer-type generator breaker (Fig. 6.7c), which is capable of interrupting 110 kA fault current and rated at 13 kA with natural air cooling or 25 kA with forced air cooling. Brown Boveri have a commerical range of HE generator breakers suitable for lower generator ratings which utilise the SF_6 self-extinguishing principle discussed in Chapter 3 and Section 7.5 (Dunckel and Vogt, 1985).

There have been discussions concerning the possibility of using the transformer-side circuit breaker for generator protection. Kulicke and Schramm (1980) report the ability of a full duoflow SF_6 breaker on the high-voltage side of the transformer (Fig. 6.7a) to meet the above demands at the Itaipu 550 kV power station, Brazil/Paraguay. Sufficient arc resistance to shorten the current-zero delay is achieved with four breaks per pole, the use of graphite nozzles/contacts (to avoid high-conductivity metallic contamination of the arc plasma) and arc lengthening by rapid arc-root movement into the nozzles.

6.4.2 High-voltage DC circuit breakers
High-voltage direct-current transmission of power is attractive for reducing transmission losses, increasing the power-transmission capability of a system, increasing the stability of AC links, and for linking asynchronous AC systems. Existing systems (e.g. Cabora Bassa, Inga-Shaba, Nelson River, cross-Channel systems) are of a 2-terminal variety which can operate reliably without a DC circuit breaker, since faults can be controlled by the convertors and circuit

breakers on the AC side of the convertors. Nonetheless HV DC breakers could be useful for improving system performance and when feeding parallel lines, or when parallel-connected convertors feed the same line (Fig. 6.8a; Vithayathil et al., 1985). Multi-terminal systems are expected in the future as systems grow, and these will need HV DC breakers in order to isolate sections so that the remaining lines continue to transmit, to isolate a faulted line with minimal system disturbance and for clearing terminal or inverter station faults. However, DC circuit breakers equivalent in reliability and cost to AC breakers are only now being developed (Yanabu et al., 1981).

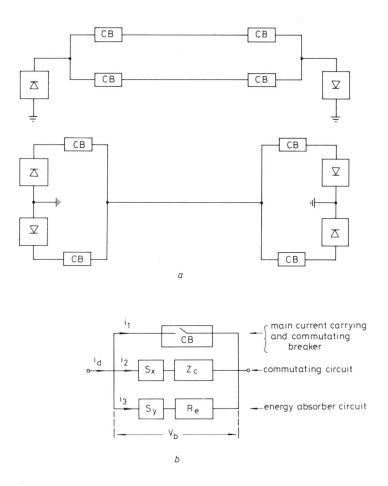

Fig. 6.8 HVDC circuit breakers
 a HVDC breakers in 2-terminal systems (Fig. 1, Vithayathil et al., 1985. Copyright IEEE)
 b Generalised circuit arrangement for DC breakers (Fig. 2, Vithayathil et al., 1983)

Impact of SF_6 technology upon transmission switchgear 109

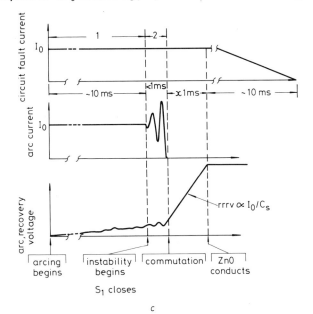

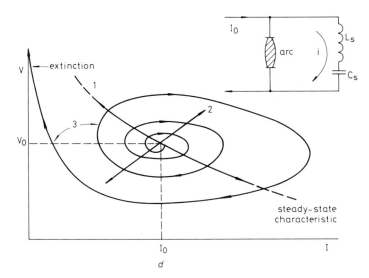

Fig. 6.8 *(continued)*
 c Phenomenology associated with DC interruption *(Fig. 2, Lee et al., 1985. Copyright IEEE)*
 d Physical description of the arc-circuit instability *(Fig. 5, Lee et al., 1985. Copyright IEEE)*

The problem in interrupting high-voltage DC systems is concerned with the greater difficulty in reducing the current to zero and in absorbing significantly greater amounts of energy than in the case of severe asymmetry of an alternating-current waveform. Although insertion of a circuit-breaker arc (as in the

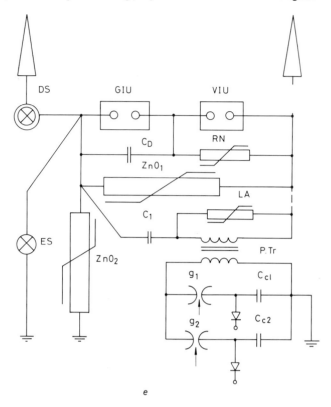

Fig. 6.8 *(continued)*
e Circuit configuration 250 kV 1200 A prototype hybrid breaker (Fig. 11, Senda et al., 1983. Copyright IEEE)

asymmetric-current case) has been sucessfully used in low-voltage systems, in high-voltage systems, the significant inductance L (due to smoothing reactors and long transmission lines) make the system time constant $\alpha X/R$ long and the energy E_b be absorbed (e.g. Vithayathil, 1983) high

$$E_b = \tfrac{1}{2} L i_d^2 V_b (V_b - V_d)^{-1} \tag{6.6}$$

where i_d = fault current, V_d = source voltage, V_b = arc voltage. It is therefore not practical to develop a circuit breaker for such systems which relies entirely upon the arc to deal with the high-voltage and energy-dissipation requirements.

Although several methods of interrupting high-voltage DC circuits have been proposed (e.g. Vithayathil, 1983), they all rely upon substantially similar operating principles which are explicable in terms of the single circuit given in Fig. 6.8*b*. Current initially flows through the circuit breaker, which is opened (Fig. 6.8*c*; Lee *et al.*, 1985) to interrupt a fault current. At a later time a commutating impedance Z_c is connected across the circuit-breaker arc by activating the device S_x. The purpose of the commutating impedance Z_c is to create a current-zero

Impact of SF$_6$ technology upon transmission switchgear   111

condition in the circuit breaker, to allow complete recovery by transferring the main current to Z_c. This commutation process varies in nature and duration with different circuit-breaking systems. The direct current flowing through Z_c builds up a high voltage V_b across the circuit breaker. When V_b reaches a value just short of the circuit-breaker rating, the energy absorber R_e is introduced through the switching device S_y and the current diverted from Z_y and the current diverted from Z_c to R_e. The use of zinc-oxide arresters (Section 6.3.2) as the element R_e eliminates the need for S_y. Therefore the circuit-breaker voltage V_b brings the main circuit current to zero according to Vithayathil (1983).

$$t_i = Li_d(V_b - V_d)^{-1} \tag{6.7}$$

If necessary, series-connected fast isolators may be opened after the current is reduced to zero to enhance the voltage-withstand capabilities. Also the energy absorbers could be connected from the breaker terminals to ground, with or without surge arresters across the breaker.

Systems based on the general layout shown on Fig. 6.8b may be further subdivided according to the nature of Z_c. The use of glow-discharge tubes for Z_c allows the commutated current to be interrupted electromagnetically (Hofmann et al., 1976). Alternatively an uncharged (e.g. Lips, 1980; Lee, 1982; Senda, 1983) or a charged capacitor (e.g. Senda, 1983; Greenwood, 1972; Yanabu et al., 1981; Lee et al; 1985; Tokuyama et al., 1985) may be used to force the circuit-breaker current to zero. the use of the uncharged-capacitor method has the advantage of simplicity, but suffers from the need for large capacitances to interrupt high currents since the interruptible current is proprtional to the square root of Z_c (Senda et al., 1983). The charged-capacitor method only requires a relatively small capacitance, but has the disadvantage of needing the capacitor to be pre-charged and triggered by less reliable trigger gaps (Senda et al., 1983).

Both these methods utilise the negative current/voltage characteristic of the interrupter arc (Fig. 6.8d) to develop a growing oscillatory current between the arc and capacitor (Fig. 6.8c). This occurs since a small increase in arc voltage forces current into the parallel capacitor Z_c, with a consequent reduction in the arc current, which, in turn, induces an additional rise in arc voltage beause of the negative arc characteristic. This unstable situation ultimately leads to the transfer of all the current to the capacitor. In the presence of stray or added inductance, the behaviour is modified depending upon the relative magnitude of the arc time constant τ and the angular frequency of the arc, C_s and L_s network (Fig. 6.8d). A stability analysis (Lee, 1982) shows that, if the arc time constant is less than a critical value τ_c

$$\tau < \tau_c = -(\tfrac{1}{2})rC_s + [(\tfrac{1}{4})r^2C_s^2 - (r/R_0)L_sC_s]^{1/2} \tag{6.8}$$

(R_0 = DC arc resistance) and that the dynamic arc resistance r is negative, then the oscillatory current will grow owing to the arc moving along the spiral locus on Fig. 6.8d. In order to ensure that $\tau < \tau_c$, τ_c needs to be made as large as practical. The most economic way of achieving this is to design the interrupter to produce a large dynamic arc resistance r. Also insertion of a contrary-charged capacitor accelerates the onset of instability, since the arc time constant decreases with current.

Because of the large number of possibilities that exist in evolving particular circuit-breaking systems from the general layout shown in Fig. 6.8b, and because consideration also needs to be given to the protective role of breakers on the AC side of the DC convertors, it is difficult at present to specify uniquely the requirements for the interrupter unit. Nonetheless it is apparent that four major demands need to be considered, relating to voltage-withstand, fault-current, energy-absorption and switching-time capabilities. The required energy capability is specified by eqn. 6.6 and the switching time by eqn. 6.7. As a result, three possible categories of interrupters may be identified. These are: a fast-acting (30–50 ms) full-voltage high-interrupting-current unit; a full-voltage load-current interrupter with a total operating time of 40–100 ms; and a slow-acting (100–200 ms) low-voltage load-current interrupter (Vithayathil 1983).

Within this general framework there is no overall preference for a particular interrupting medium, and indeed vacuum, oil, air-blast and SF_6 have all been used in various systems (Vithayathil, 1983). However, in the ETNA–EDF breaker (Gabin et al., 1974) the arc is quenched by intense cooling with hypercritical SF_6 maintained at very high pressure. This type of breaker was tested at the Echinghen convertor station of the cross-Channel DC link to interrupt direct currents up to 1 kA, with a peak breaker voltage of 190 kV against a maximum rectifier source voltage of 120 kV.

The Hughes breaker (Hofmann et al., 1976) utilises a high-speed (opening time, 1·5 ms) SF_6 interrupter (for carrying the main current and to generate a 400 V arc) in conjunction with electromagnetically controlled glow-discharge tubes. This type of breaker has been tested on the Pacific Intertie in the USA, and interrupted about 800 A with a peak voltage of about 93 kV across the breaker.

The Toshiba slow-acting breaker (Yanabu et al., 1981; Senda et al., 1983) utilises a hybrid interruption unit, whereby vacuum and Sf_6 interrupters are series-connected with a voltage-dividing RC circuit connected in parallel (Fig. 6.8e) (Senda, 1983). This is intended to combine the high di/dt versus dV/dt capability of the vacuum interrupter, for the initial voltage recovery, with the better dielectric properties of the SF_6 interrupter for the later stages of the recovery. A full gas-insulated prototype breaker rated at 250 kV 1200 A continuous voltage and current, and 420 kV peak recovery voltage and 1·44 kA interrupting current, has been developed (Senda, 1983).

Recently, systems using SF_6 puffers as the sole interrupting unit have been developed, one utilising a charged capacitor for Z_c (Fig. 6.8b) (Tokuyama et al., 1985) and the other utilising an uncharged capacitor (Lee et al., 1985).

Lee et al., (1985) have modified a conventional AC puffer breaker in order to maximise the dynamic arc resistance r in accordance with the requirements of

Impact of SF_6 technology upon transmission switchgear   113

eqn. 6.8 to reduce the arc time constant below the critical value for oscilliatory behaviour. These modifications involved increasing the interrupter stroke by 14% to give longer arcs. In conjunction with an increase of 15% in piston diameter and a 20% increase in opening speed, this also produced a higher transient pressure. In addition, the nozzle-throat diameter was optimised and a double-flow system utilised (Lee *et al.*, 1985).

The complete circuit breaker utilised four series breaks, each paralleled with a ZnO arrester and capacitor (Lee *et al.*, 1985) which were inserted into the circuit by vacuum switches S_I. Closing resistors R were inserted with switches S_2. The breaker has already been field tested on the Pacific Intertie at Celeilo, Oregon, and shown to be capable of interrupting a fault current of 1·2 kA at 400 kV and to withstand a maximum voltage of 700 kV. To ensure commercial viability, the breaker is of modular design, adaptable for different DC systems and built around standard production-type components.

6.5 Particular examples of SF_6-insulated installations

The development of SF_6-insulated metalclad systems has contributed in a majoi manner to a number of ambitious transmission-substation installations. The compact, totally enclosed and versatile nature of such systems has opened up new possibilities for installations in urban areas, in main transformer substations, in difficult industrial environments, in underground power stations and in hybrid substation systems.

Space or environmental considerations increasingly prohibit the installation of open conventional systems in urban areas, but both constraints can be met by the 90% reduction in space permitted by metal clad systems and the possibility of their economic installation indoors. Such achievements are particularly possible with systems operating at voltages between 72 and 145 kV, because all three phases of the supply can be accommodated within a single SF_6 enclosure (Section 6.3.1). Also, maximum economy is achieved when transformation from the transmission voltages of 170/245/420/525 kV down to the distribution voltage takes place as close as possible to the load centres, which are normally in urban areas. with SF_6 gas-insulated systems there is usually an unlimited choice of sites. Examples of the layouts for typical 420 kV substations are given in Szeute-Vargas and Tacchio, 1983, and Ali and Headley, 1984.

Operational reliability and space requirements are also factors influencing the choice of equipment for electricity-supply installations in large industrial plants such as steelworks, aluminium refining plants etc. Often, there is a high level of pollution associated with such plants, which can create serious problems with high-voltage insulators and necessitate the use of special insulator-washing equipment. Corrosion of conductors and fittings may also arise. SF_6-insulated systems and switchgear are not influenced by industrial pollution, so that no additional maintenance is required and there is no associated threat to reliability.

Connections can be made directly to the transformers, and incoming high-voltage lines are terminated at a safe distance from the site for interconnections by SF_6-filled tubular busbars. In this manner, exposure of conductors and high-voltage insulators to pollution can be further reduced. An example of such a system, the 362 kV SF_6 substation in Warri steelworks, Nigeria, is given by Szeute-Vargas and Tacchio, 1983.

The advent of SF_6 switchgear technology has enabled entirely new methods to be used in the planning and installation of thermal and large-hydroelectric power stations and of underground power stations. In the case of thermal and hydroelectric power generation, the improved insulation level allows the switchgear to be coupled directly to the energy source, thereby permitting a space-saving arrangement for the electromechanical chain consisting of generator, transformer and switchgear. All the protective control and supervising equipment can be arranged with the shortest cable routes to improve plant reliability. A most impressive example of such an installation is the 525 kV SF_6 Itaipu hydroelectric power plant, Brazil/Paraguay (Szeute-Vargas and Tacchio, 1983). Instead of building a conventional outdoor station at a considerable distance from the dam, the SF_6 switchgear was installed directly above the transformers, between the generator hall and the dam (Szente-Varga and Tecchio, 1983). The compact and enclosed nature of SF_6 switchgear also makes it particularly suitable for underground installations in caverns with their own particular ambient conditions. Examples of such installations are the 2-break 420 kV GEC metal clad circuit breakers for the CEGB pumped-storage station at Dinorwig in N. Wales, which are accommodated, for environmental reasons, within caverns between the two water levels used in the scheme. The use of vertical oil-filled cables for the outgoing high-voltage connections gives rise to problems concerning the static oil pressure once the difference in height exceeds a certain value. These problems can be overcome with SF_6-insulated connections as used in the 420 kV Drakensberg pumped-storage installation, which utilises a vertical SF_6-filled connection to the overhead transmission lines.

Since air-insulated substation assemblies are already in existence, the possibility exists for extending such arrangements owing to increased supply demand etc., by installing a supplementary SF_6 system. Such hybrid installations, involving two insulating media (air and SF_6) having different characteristics, require careful insulation co-ordination, as already indicated by the considerations made in Section 6.3.2 (Andersen and Simms, 1983). Two basic types of hybrid installations may be distinguished; namely, installations in which only the busbars and busbar isolators employ gas-insulated technology, and installations in which the busbar and busbar isolators are of conventional design with air insulation, but all the other equipment consists of SF_6-insulated components. Where space is at a premium, the former installation is chosen, as is the case for the 245 kV hybrid installation at Galmiz (Lausanne, Switzerland). Extension of an existing air-insulated installation, without utilising additional land, leaves little alternative but to use SF_6-insulated systems. An example of such an exercise is the upgrading of the 420 kV installation of Elektrizitats–Gesellschoft

Impact of SF_6 technology upon transmission switchgear

Laufenburg AG (which forms an important junction interconnection between the high-voltage systems of France, the Federal Republic of Germany and Switzerland) from a double- to a triple busbar system with provision for future conversion to a quadruple system (Crameri *et al.*, 1982).

The Electricity Supply Commission of South Africa is installing a 765 kV grid whose basic insulation level of 2400 kV is determined by parts of the system being sited at an altitude of 1800 m (Lohmann and Bolton 1985). Conventional switchgear for this system would need to be rated for a service voltage of 1000 kV, and as such would need to be specially developed. SF_6-insulated systems can be more easily uprated and, additionally, the insulation level is unaffected by altitude. For this system, a 550 kV gas-insulated system is being uprated with enlarged enclosures diameters.

Chapter 7

Impact of SF$_6$ technology upon distribution and utility switchgear

The distribution function may be regarded as consisting of two stages. The first stage is used to bring the medium-voltage supplies to the centres of use, whilst the second stage conveys the medium-voltage supply to the centres where it is converted to a low voltage for industrial or domestic use. Typically, although not exclusively, primary distribution voltage levels lie in the range 66–145 kV, whilst secondary distribution may be at 3.6–36 kV. The precise levels may vary from country to country and also from region to region within a country. Although 11 kV is usual for secondary distribution in the UK, the North Eastern Electricity Board has a substantial distribution at 20 kV (Bartle and Gowett, 1984).

The switchgear for these purposes in the past have traditionally utilised either insulating oil or air as the arc-quenching media, whereas more recently vacuum or SF$_6$ is being increasingly used. In order to appreciate the impact being made by SF$_6$ technology upon distribution switchgear, it is necessary first to discuss the requirements imposed upon the circuit breaker.

7.1 Operation and system requirements

Requirments demanded of distribution-type switchgear are governed by two major factors. The first is the performance requirement determined by the system parameters; e.g. lumped and distribution impedances, fault levels, phase differences in interconnected systems etc. The second is the operational requirements determined by specific applications; e.g. frequency of metalclad construction, integral earthing facilities etc.

Operational performance requirements include compliance with national or international specifications for type testing, but other requirements can differ considerably depending upon whether the circuit breaker is used for primary distribution or for specialised secondary duty, e.g. for controlling arc furnaces. Furthermore, in the UK it has been the practice for all switchgear-earthing operations to be safe with positive interlocking, and this has involved circuit

Impact of SF_6 technology upon distribution and utility switchgear 117

breakers up to 36 kV being withdrawable, giving isolation from both busbar and circuit. In the mining, steelmaking and petrochemical industries one outage due to the non-withdrawable nature of the switchgear can cost more in terms of lost output than the initial cost of the switchgear (Headley, 1984). A reduction in the amount of maintenance may well be desirable for applications involving a large number of operations, but clearly may not be so critical for situations involving only a small number. For primary distribution the switching duty relative to circuit breaker-life is not onerous [6 low-level faults per annum on auto re-close, one operation every few years on urban systems (Walker, 1984)], but in industrial-plant applications there may well be requirements for frequent ($\sim$ thousands operations per year) high-rated switching duties. Although flammability has not been a serious problem in the UK, (e.g. 250 000 switches, 75 000 circuit breakers using 3·5 million gallons of oil are estimated to exist in the UK at present (Headley, 1984; Oakes, 1986)) the use of oil is prohibited in high-rise buildings, petrochemical plants etc., and export trends indicate the need for alternatives.

Although most switching-performance requirements are well defined and can be covered by type testing (e.g. rated short-circuit making and breaking capacity) some are not so easily defined (e.g. low inductive current switching). Ratings associated with high inductive current switching are well defined and are similar for the new (vacuum and SF_6) and traditional technologies (bulk oil) (Headley 1984). However low inductive switching is prone to 'current chopping (current interruption not coincident with the zero of current associated with the power frequency and caused by arc instability) or high-frequency effects in the circuit being switched. The voltage peak resulting from such effects is given approximately by, for example, Headley (1984):

$$V = \beta I_{\text{chop}} Z_{\text{effect}} \qquad (7.1.)$$

β is a parameter determined by the extent of retrieval of energy stored in the system inductively. Eqn. 7.1 shows that overvoltages are generated when high currents are chopped (I_{chop}) in circuits with a high surge impedance Z_{effect}. Since current chopping is related to arc instabilities, the potential overvoltages developed will, in principle, be influenced by the arcing environment, and hence the type of curcuit breaker. the switching of transformers feeding a significant secondary load is usually not a problem since arc instability is unlikely at high currents, and anyway the energy in the transformer inductance would be dissipated in the secondary with no overvoltages. Neither does the switching of transformers with no secondary load usually cause problems, since the steady-state no-load magnetising current is too low (for an 11 kV 2·5 MVA transformer the magnetising current is about 2 A). However, under current in-rush conditions, higher current levels are involved but the arc becomes more stable and less prone to produce current chopping. For transformer switching the value of β in eqn. 7.1 is typically 0·45 owing to hysteresis effects (Headley 1984). Although the switching of motors under load or no-load conditions is not a problem (since only small voltage disturbances are produced due to the running motor continuing to supply its own magnetising current), the disconnection of motors with a stalled or locked rotor can cause current chopping similar to the no-load transformer switching, since no back EMF is generated.

Certain combinations of circuit elements can result in high-frequency current discharges during making operations or as a result of arc reignitions during opening operations. The interruption of the currents in the presence of high surge impedances, such as those involved in arc-furnace operation (involving the combination of power-factor-correction capacitors, a highly inductive transformer and the circuit breaker), can produce excessive overvoltages. Although the capacitive-current switching ability of circuit breakers is well defined by equipment manufacturers, problems can arise with the switching of capacitor banks with little or no interconnection impedance.

7.2 Relative merits of SF_6, vacuum and more traditional circuit breakers

The increase in the proportion of the switchgear market being captured by the new technologies (vacuum, SF_6) (Fig. 7.1a,b) suggests that they are better able to meet the general requirements of distribution systems as discussed above. In the UK the new technologies' share of the 12 kV market has increased from 5% in the decade upto 1984. SF_6 interrupters for distribution systems have only been available in the UK since about 1981, so that their impact is delayed compared with that of vacuum. In general, UK industry has responded more rapidly to the new interrupters than the utilities mainly on account of the new technologies being better able to meet the more onerous switching demands of private industry (more frequent switching operations and dereased maintenance requirements) than the traditional circuit breakers.

Although the general trend in the European Community (Fig. 7.1b (ii)) is similar to the UK in replacing the traditional circuit breakers by vacuum and SF_6, the SF_6 share of the market is greater in Europe.

Traditional distribution switchgear suffers from a number of potential disadvantages. These include the need for periodic maintenance due to the deterioration of the insulation oil and contacts, and the need to examine the interrupter after each fault operation and after a modest number of operations on load. With vacuum and SF_6, the trend is towards a sealed-for-life concept. Gas leakage in service is virtually non-existent with both vacuum and SF_6, but, in the latter case, gas pressure can be monitored and pressure switches can initiate tripping or signal a warning in the event of a pressure fall (Blower, 1984; Lister, 1984).

Vacuum and SF_6 interrupters are generally less susceptible to current chopping than oil and air interrupters (Fig. 7.1c).

In the case of vacuum interrupters chopping occurs for currents in the range 0·1–20 A (e.g. CIGRE, 1980, 1981) and pre- and restriking can occur during

Impact of SF$_6$ technology upon distribution and utility switchgear 119

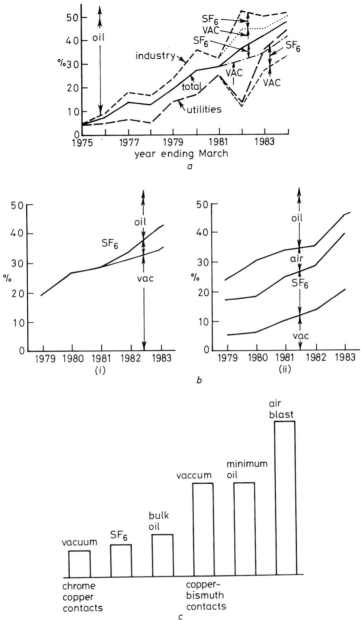

Fig. 7.1 Comparisons between oil, air, vacuum and SF$_6$
 a Growth of new technology in UK 12 kV distribution switchgear market (Fig. 5, Blower, 1984a. Copyright IEEE)
 b Impact of new technology on 24 kV circuit breakers (Fig. 1, Blower, 1984b)
 (i) UK
 (ii) EEC
 c Relative current-chopping ability (Fig. 4, Headley, 1984b)

closing and opening owing to the rapid recovery of dielectric strength ($< 10\,\mu s$ (Blower *et al.*, 1978)). Virtual current chopping can occur when a disturbance initiated by chopping in one phase is coupled electromagnetically to a second phase, causing premature current chopping in the second phase (e.g. Lister, 1984). Although such effects may be reduced through the use of low-thermal-conductivity and higher-vapour-pressure contact materials, this is achieved at the expense of high interrupting ability.

In the case of SF_6 interrupters there is increasing evidence that rotary-arc and self-pressurising interrupters are completely free from such current-chopping limitations (Parry, 1984) on account of their interrupting characteristics being governed by the current itself.

Although the vacuum interrupter has much to commend it, there are a number of considerations which can, nonetheless, make SF_6 preferable. For the circuit-breaker manufacturer, the vacuum interrupter is available as a finished component with an established rating which only requires a driving mechanism to convert it into a circuit breaker (Blower, 1984). Unfortunately different interrupter manufacturers achieve different compromises between the factors which govern performance; e.g. the choice of contact material which affects susceptibility to current chopping. Consequently, the circuit-breaker manufacturer is susceptible to the need to produce a variety of mechanisms to cater for the different characteristics of the interruter available from the different manufacturers (Blower, 1984). For the circuit-breaker purchaser, reassurances are often sought from the manufacturer (who is susceptible to the above limitations) concerning the production of overvoltages, the monitoring of the vacuum level and the air insulation of the live parts of the circuit breaker.

With SF_6, more of the end product is under the control of the circuit-breaker manufacturer, so that for countries with no indigenous source of vacuum interrupters (e.g. France), this consideration alone may dominate. Furthermore, the circuit-breaker manufacturer has a choice of the principle of operation [puffer, self-pressurising, rotary arc (Chapter 3)], which may be assisted with the information presented above. In the case of the puffer breaker, the energy requirement of the mechanical drive is of paramount importance. In practice, this energy is not significantly different from that already used for traditional circuit breakers. Furthermore, heavy contact loads are required for vacuum circuit breakers to close quickly onto faults on account of the butt nature of the vacuum-interrupter contacts. The quest by circuit-breaker designers for a reduction in the mechanical-energy requirements has been met with SF_6 interrupters using the self-pressurising and rotary-arc principles. However, any form of circuit breaker using the fault current itself to assist arc extinction is susceptible to the problem of reduced effectiveness at low fault currents, allied with the danger of becoming excessive when the fault current is high. The mechanical and electrical endurance of SF_6 switchgear is such that internal overhaul is unneccessary during economic service life in all except the most onerous system requirements; e.g. arc-furnace control (Blower, 1984).

Problems associated with the chemical reactivity of the arcing by-products,

Impact of SF_6 technology upon distribution and utility switchgear

which were encountered during the early development era of SF_6 interrupters, have now been totally overcome through the use of appropriate materials and proper quality-control procedures (Chapter 9). Fears concerning the toxicity of these by-products (Section 9.1.3) also need to be placed in perspective. Thus the mass of gas used in such SF_6 circuit breakers is small and the quantity of arcing by-products even smaller, provided prolonged arcing due to internal flashover is avoided. There is therefore a need to design the circuit breaker which operates with a low nominal pressure and reduced risk of internal flashover. The practice of using phase-separated components in a housing of insulated material and encapsulated connections (e.g. Blower, 1984) to reduce the risk of flashover would seem to be preferable to constructing the unit in such a way that any by-products are directed away from places where human operators might be located, should an internal explosion occur. Also, self-generating interrupters require pressures of 3 to 8 atmospheres, depending upon rating, whereas the puffer type of interrupter can operate with nominal pressures of about 1 atmosphere.

7.3 Puffer circuit breakers

SF_6 puffer circuit breakers for distribution systems have been developed by various manufacturers, often using the considerations already discussed in Chapter 4. Because of the lower voltages involved at distribution levels, higher temperatures of the arc-quenching gases can be tolerated, with a consequent reduction in the volume of the piston chamber. The use of separate arc piston chambers can also lead to more efficient performance (Chapter 4 and Ueda *et al.*, 1982).

Mitsubishi have produced a 7·2 kV 63 kA puffer breaker operating with a nominal pressure of $5\,\text{kg cm}^{-2}$ gauge. The three phases are housed seperately, but mounted together on a single console. The breaker's interrupting chamber is constructed from aluminium casting with epoxy-resin bushing.

Yorkshire Switchgear produce a puffer breaker rated at 31·5 kA at 12 kV and 25 kA at 24 kV using a nominal gas pressure no greater than 1 bar gauge (Blower, 1984). The design involves a one-piece resin moulding, which incorporates three separate cylinders in which the contacts and their operating pistons are located. In this way the three phases are enclosed within the same unit, but, at the same time, the risk of internal flashover is considerably reduced. The common housing of phases in this manner can, of course, be achieved economically for secondary-distribution breakers, as opposed to transmission breakers, on account of the considerably lower voltages involved. Gas leakage is also minimised by reducing the number of joints to three (between metal chamber and insulated moulding, the mechanical drive shaft and the filling orifice). Furthermore, the live conductors are totally encapsulated to reduce environmental effects.

7.4 Rotary-arc circuit breakers

Various types of rotary-arc circuit breakers, which do not deliberately utilise auxiliary assistance in the form of nozzle flow, are commercially available for the approximate voltage range 3·6–36 kV and currents upto 50 kA. South Wales Switchgear, Brush Switchgear and Yaskawa utilise the helical-arc type of interrupter (Fig. 3.3a) with units covering the range 3·6 kV/12·5–50 kA, 7·2 kV/12·5–40 kA (Yashawa; 3·6 kV/12·5 kA, 12 kV/25 kA upto 2000 A nominal current, 36 kV/25 kA (South Wales Switchgear; Brush Switchgear). Merlin–Gerin utilise arc rotation along concentric ring contacts to cover a similar voltage range.

The relative merits of the helical-arc and concentric-ring contact interrupters are not well documented, but the similar performance of each type would suggest that there are no outstanding advantages of either for present applications. Both types utilise main and arcing contacts. The main contacts not only carry the rated current when the circuit breaker is closed, but also serve to short-circuit the magnetic-field-producing coil (e.g. Fig. 7.2a). The opening sequence requires that this main contact opens first, diverting the current through a subsidiary pair of contacts. When these open, an arc is formed (Fig. 7.2a) and, as the moving contact passes the main ring contacts, arc transfer occurs accompanied by a current flow through the coil. The resulting build-up of magnetic field rotates the arc between the ring contacts.

The transfer of the arc to the main arcing contacts is encouraged by a suitable design of the magnetic circuit. There is evidence (Zhang *et al.*, 1985) that the transfer time is of the order of 3 ms, being insensitive to gas pressure in the range 1–5 bar gauge, current levels of 1–10 kA and a transfer gap of 3–6 mm. The dominating feature would appear to be associated with the growth of the magnetic field due to current flow through the coil, which is initially transient in nature (Spencer, private communication; Turner and Chen, 1984). There is also evidence (Zhang *et al.*, 1985) that, under some operating conditions, the arc voltage can increase to such an extent that a restrike can occur across the transfer gap. This essentially forms a parallel conducting path across the coil with a consequent reduction in the driving magnetic field.

An alternative form of contact separation is used in the Merlin–Gerin 10 ka, 7·2 kV 'Rollarc contactor'. The main current contacts are concentric with the ring arcing contacts, but are separated before the ring contacts in order to activate the magnetic-field-producing coil. This allows a rapid acceleration of the arc rotation, without the need for arc transfer. As a result, the contactor is rapid acting and eliminates voltage surges (Duplay and Hennebert, 1983).

Current chopping and the associated overvoltages, which have proved troublesome with vacuum interrupters, are claimed by manufacturers to be absent throughout the complete range of operation owing to the relatively high resistance of the spiral arc coupled with the current-dependent nature of the arc control (Parry, 1984). It has not, in practice, been necessary to use a supplementary gas flow and its associated complexities in order to interrupt low currents

Impact of SF_6 technology upon distribution and utility switchgear 123

when the electromagnetic driving forces are minimal. Neither is flow assistance necessary to remove arc by-products from the contact gap in order to meet recovery-voltage demands, since this would appear to be achieved purely by the rotational motion of the arc in the concentric-ring arrangement and, additionally, by the pumping action of the alternate elongation and short-circuiting of the spiral arc column in the helical-arc arrangement. The amount of energy for operating these interrupters is small, and the closing-energy requirement is approximately half that required by the equivalent oil circuit breaker (Parry, 1984). A consistently low duration of arcing ($\sim$ 1 cycle at 11 kV, 3·5 half-cycles at 36 kV) over a wide range of current values (e.g. Yaskawa up to 50 kA at 3·6 and 7·2 kV; South Wales Switchgear at 12·5 kA, 13·8 kV;), combined with the rotation of the arc root, leads to reduced contact wear. Although the number of successful operations decreases with current level these remain nonetheless substantial [> 300 operations at 6 kA (Parry, 1984] and in excess of normal requirements for both distribution supply and industrial control. In all these respects, the rotary-arc-interrupter performance exceeds present supply-network requirements, so that the sealed-for-life maintenance-free claim made by the manufactuters would appear to be justified.

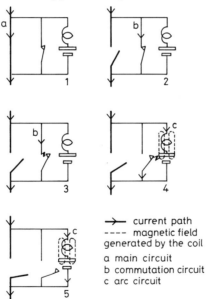

Fig. 7.2 *Rotary-arc circuit breakers Opening sequences Using arc transfer (Fig. 25, Duplay and Hennerbert 1983)*
  c (Merlin–Gerin)

For 3-phase operation at distribution level, the possibility exists of enclosing the interrupters associated with each phase within a common gas-filled enclosure, on account of the lower voltage levels involved than with transmission switchgear. The relative disposition of gaps requires careful consideration to prevent not only interference with the electrical recovery of each phase by hot gases expelled from the other phases, but also deleterious interaction of the driving magnetic forces from each phase. The geometric layout of the contact gaps associated with each phase depends upon the manufacturer: For instance, South Wales Switchgear utilise a trefoil arrangement of coils for their lower-rating breakers, where space availability is dictated by minimum-volume requirements demanded for use in restricted locations, such as underground mining. For higher nominal current ratings (1·25 and 2 kA), the volume of existing gas enclosures is greater, so that it is possible to arrange the field coils in line, resulting in a further simplicfication of the mechanical-drive geometry.

Typical pressure levels used in rotary-arc circuit breakers are 3 bar. The breakers are tested to show that the pressure will not fall below the certification level after at least 10 years in service. Nonetheless, the circuit-breaker insulation is maintained even if the pressure decreases to atmospheric and the rated normal current can be safely interrupted at this pressure. Furthermore, a pressure-operated device operates a trip, or inhibits tripping, as required, should the pressure fall to this level.

7.5 SF_6 Self-extinguishing circuit breakers

The most sophisticated form of self-extinguishing circuit breakers to date utilises arc rotation for gas heating, as already indicated in Fig. 3.4a, to encourage uniform heating, which is claimed to be important (Zimmermann, 1984). The self-pressuring breaker offers a number of advantages over the conventional puffer breaker at distribution-voltage levels. First, the energy required to generate the necessary gas pressure with the self-pressurising breaker, in order to produce the circuit-breaker performance, increases more than proportionally with current (Fig. 7.3; Zimmermann, 1984), reaching about 7 kJ at 100 kA. For 63 kA fault current, the self-pressurising breaker only requires about 10% of the operating energy needed by a corresponding puffer breaker. Moreover, this relationship is practically independent of the rated voltage of the circuit breaker.

Secondly, although an auxiliary puffer is utilised for the interruption of small currents, the mass of gas which has to be moved by the auxiliary piston is less than that in a conventional puffer, so that a simpler and lighter mechanism can be used with a reduction in the driving-energy requirements.

Thirdly, whereas puffer circuit breakers normally require a high contact speed in order to generate the initial compression, with the self-extinguishing principle, the contact speed is determined only by the rated voltage (contact clearance) and the required operating time.

Fourthly, the self extinguishing breaker retains its interrupting capacity even after the contacts have reached the fully open position. This favourably influences the dimensions of the interrupter, since no special measures are required to

Impact of SF$_6$ technology upon distribution and utility switchgear 125

accommodate the longer arcing times arising from short-circuit currents with delayed current zeros.

Fifthly, since the self-pressurisation decreases with fault current, the breaker has a soft interrupting characteristic which prevents precurrent zero-waveform distortion. Thus the small inductive currents associated with switching motors and unloaded transformers do not promote dangerous overvoltages, and the need for overvoltage-suppression equipment is avoided (e.g. Satyanarayana and Braun, 1984). Of course, both rotary-arc and puffer breakers can also be designed to have soft interrupting characteristics.

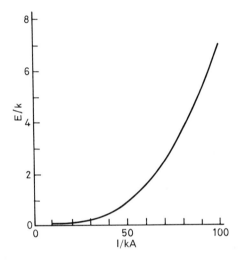

Fig. 7.3 *Self extinguishing breaker: energy required to provide sufficient pressurisation for arc interruption (Fig. 1, Zimmermann, 1984)*

Self extinguishing SF$_6$ circuit breakers have been in service since 1977. To date there are more than 7000 units alone of the Brown Boveri type HB operating worldwide (Zimmerman, 1984). Service experience has shown that these circuit breakers are still completely serviceable even after 30 000 close/open operations without maintenance. On-site tests with 0·1–9 MW motors have yielded a maximum overvoltage factor of 2·1, so confirming the soft interruption characteristic of the circuit breaker (Zimmermann, 1984). Brown Boveri have recently designed a type HE 100 kA 24 kV self-extinguishing breaker for application as a generator circuit breaker, as well as for the switching of voltage-compensating equipment, high-current distribution and very large motors.

Although use of the self-extinguishing principle is subject to the same physical limitations as other switching principles, there is evidence that neither in relation to rated voltage nor interrupting capability have the limits been reached (Zimmermann, 1984). Although the voltage limits are likely to be determined by the temperature dependence of the dielectric strength of SF$_6$, it is expected that the principle may well be applicable for voltages of 72·5 kV and above. However, the interruption of currents in excess of 100 kA is likely to lead to limited RRRV capabilities unless ohmic or capacitive damping elements are employed.

7.6 Insulation of distribution switchgear

Insulation developments related to distribution switchgear have involved improved insulation of peripherals and connections on to the circuit breaker. Two main avenues of evolution may be identified following the introduction of SF_6. The first is an extension of the gas-filled-system concept used at transmission-voltage levels, and the second makes more extensive use of solid insulation on the circuit-breaker itself. Both approaches are designed to exclude the influence of the open environment from the vicinity of the circuit breaker, which under many circumstances is not only desirable but necessary. For example, in tropical climates humidity levels of 90% are commonplace, so that, when rapid cooling of humid air occurs during a tropical rainstorm, droplet formation is induced. Under such conditions, insulation which is free from external discharge under normal (dry) conditions may well suffer surface discharges, which in time may degrade the insulation level (Oakes, 1984).

As in the transmission-voltage systems (Chapter 6), the use of gas-filled insulation is also intended to provide more compact substation packages with the associated economic benefits. However, at distribution voltage levels there is an additional potential benefit in that power distribution through an existing substation can be increased by raising the distribution-voltage level whilst maintaining the same volume of the switchgear and peripherals due to the better insulation strength of SF_6. The options available with such gas insulation at distribution or subtransmission levels include a choice of high- or low-pressure SF_6, single or double O-ring sealing arrangements, single- or 3-phase enclosures, and aluminium or steel as enclosure materials (Wieland *et al.*, 1982). Whereas at transmission-voltage levels SF_6 pressures of 2·5–5 bar are used, at distribution-voltage levels lower SF_6 pressures may be economically, as well as technically, more desirable. For instance, withstand voltages do not increase linearly with pressure since the dielectric strength becomes more sensitive to surface roughness and contamination, leading to increased manufacturing costs. Furthermore, a lower gas pressure results in a reduction in the rate of loss of dielectric strength should a specific leak occur (Wieland *et al.*, 1982). Since 98% of all leaks occur in the sealing system, it is essential to minimise the stress on the sealing gaskets. This stress s_D depends upon the total length of the gasket L_D and the difference between nominal P_N and ambient P_O pressures (Wieland *et al.*, 1982):

$$S_D = (P_N - P_0) L_D \tag{7.2}$$

Evaluation of S_D for different systems shows that a 3-phase system at relatively low nominal pressure (< 1·5 bar) has approximately only a quarter of the stress of a high-pressure (> 2·5 bar) single-phase enclosure and about one-seventh of a high-pressure 3-phase system. This has led Wieland et al. (1982) to use a low-pressure 3-phase enclosure system with a different gasket sealing system for distribution-voltage insulation. A low-pressure mixture of 95/50% SF_6/air has been used which has a somewhat higher dielectric strength than pure SF_6 (Wieland et al., 1982); this reduces greatly the possibility of liquefaction at lower temperatures and facilitates filling the enclosure without the need for prior evacuation. As a result, a cubicle-type gas-insulated system has proved more economic for the 46–72·5 kV voltage range than the pipe-type systems (Section 6.3) used at transmission-voltage levels (Wieland et al., 1982). The circuit breaker itself, which is of a spring-operated puffer type, is also enclosed within the SF_6/air insulation in a sub-compartment. The first systems of this kind were installed in 1979, and up to 1982, 200 such systems had been ordered or installed (Wieland et al., 1982). The system can accommodate single or double bus arrangements.

The SF_6-filled insulation- enclosure principle has also been applied to systems using vacuum interrupters as the circuit breakers (Noble, 1984).* Vacuum interrupters would appear to be more susceptible to corrosion in high-humidity climates since ozone produced during any surface discharges may combine with atmospheric nitrogen to form nitrogen oxides, which, in turn, react with any condensed water to form nitric acid. Long-term exposure of sensitive metallic parts, such as the vacuum-interrupter bellows, can lead to corrosive degradation (Oakes, 1984). The problem may be avoided by enclosing all live metallic parts, including the vacuum bottles, within an SF_6-filled enclosure at a pressure of less than 1 bar gauge. More careful consideration of the enclosure-pressure levels would appear to be necessary with vacuum rather than SF_6 interrupters, in order to ensure no adverse effects upon the bellows operation etc.

The second method for eliminating environmental influences by making more extensive use of solid insulation involves encapsulating all external live metal using cast-resin insulation. This allows the receptacle-transformer-bar primary bushing and disconnectable cable termination to be integral parts of a single cast-resin moulding. The main body of the SF_6 puffer breaker is a 3-phase 6-pole single-piece cast-resin moulding (Oakes, 1984). This houses three separate cylinders for accommodating each of the three phases separately, but within the same SF_6 enclosure. The contacts associated with each phase, and their operating pistons, are housed within each cylindrical volume of the moulding.

7.7 Fuse–switch combinations

Fuse–switch combinations are used for both economic reasons and to reduce the frequency and extent of consumer outages. The use of fuses with contactors or

---

* More recently, protective relays tend to be located in separate cubicles (Noble, 1987, private communication).

circuit breakers has, in the past, been primarily for the control of motors in large industrial or power-station auxiliary-switching applications. The fuse–switch combination has also been used in secondary high-voltage-distribution networks for the control and protection of transformer circuits (e.g. Pryor, 1984). However, the development of new SF_6 interrupters which are economically competitive is rapidly changing this pattern.

The switch of the switch–fuse combination is used for switching normal loads, overload and possibly low-current faults, while the fuse is used to cater for the high-current short-circuit fault conditions. The action of the fuse, switch and tripping relay must, of course, be co-ordinated, and this may be achieved by superimposing the melt-time/current characteristic of the fuse, the trip-time/current characteristic of the overload relay, and the current rating of the interrupter (Lister 1984). The crossover point between the relay and fuse characteristics represents the current above which the fuse should clear the circuit before the interrupter is operated (Lister, 1984). Some ratings for typical interrupter candidates using air, vacuum and SF_6 for motor-control applications are given by Lister which show that all these interrupters would be suitable in combination with the particular high-voltage fuse. The SF_6 interrupter in this case utilised the rotary-arc principle, and provides for the interruption of a higher overload current (10 kA) than the vacuum interrupter (4 kA). Present designs of vacuum and SF_6 contactors are typically smaller and lighter in weight than air-break interrupters (Lister, 1984). Since one duty of the contactor is to interrupt at the lower end of the current scale, susceptibility to current-chop induced overvoltages is an important consideration. The SF_6 rotary-arc interrupter is less susceptible to the production of such transients than vacuum. The use of surge capacitors or nonlinear metal-oxide resistors to overcome these overvoltage problems with the vacuum interrupter does so at the expense of unit size.

Recently Yorkshire Switchgear have developed a rotary-arc, SF_6-filled fuse (Oakes, 1986) Fig. 7.4b. In conventional fuses the choice of fuse element involves a compromise between the element's fusion energy and its influence upon arc extinction. With the new fuse these two requirements are separated, the fusion energy being controlled by the fuse element whilst arc extinction is governed by electromagnetic rotation in the SF_6†.

Secondary distribution has increasingly used outdoor-mounted fault-making, load-breaking switch disconnects in ring-feeder systems with fused switches for transformer connection. An example of such a ring system is shown in Fig. 7.4a, with a number of tee junctions feeding the secondary transformers. The fuse–switch combination in the transformer tee and the associated two switches in the ring main are normally mounted as a single module known as the 'ring-main unit' (RMV). In the UK, the unit has traditionally been filled with oil for insulation and the switches are operated in the oil so that this also forms the

† Further developments of the Dyscon Interrupter were described at CIRED, May 9187 (Oakes, private communication).

Impact of SF_6 technology upon distribution and utility switchgear 129

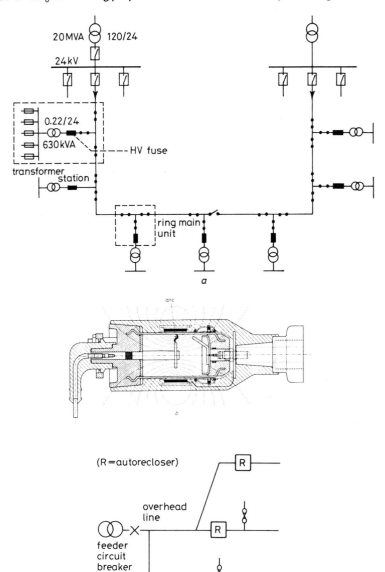

Fig. 7.4 *Fuse–switch combinations*
 a Distribution system with ring-main units (Fig. 1, Rondeel and Weissforth 1984)
 b Schematic view of Dyscon interrupter under fault conditions
 c Schematic of auto-recloser and fused network (Fig. 1, Stewart, 1984)

switching medium. The fuse could also be mounted in oil for improved insulation. European manufacturers have traditionally used air insulation and 'hard gas' devices for switching (e.g. Blower, 1984). With such ring-main units, not only does the fuse, switch and trip-relay operations need to be co-ordinated as in the motor-protection case discussed above but also the combination must be able to discriminate with the protection on the low-voltage side of the transformer.

Since 1978 the trend in Europe has been towards replacing air and oil by SF_6 as the insulating and switching interrupting medium. A Merlin–Gerin SF_6-filled indoor ring-main unit utilises three switch units sealed within a common steel tank with the fuses mounted externally. More recently, primary-distribution and motor-protection switchgear techniques have led to the transformer tee switch being replaced by circuit interrupters with respectable fault-current ratings up to about 6 kA with vacuum, and 10–12 kA with SF_6. For instance, Yorkshire Switchgear produce an SF_6 unit which utilises the puffer-interrupter principle with reduced mechanical-drive-demands. The associated fuse is accommodated externally to the SF_6 housing* using watertight elastomeric boots fitted with conductive skins to insulate the fuse and caps, and simultaneously protect against environmental hazards.

In the past, the principal advantages of the fuse over the circuit breaker has been its cheapness and its small let-through energy. However, recent advances in electronic detection methods enable very short total break times to be achieved with SF_6 and vacuum circuit breakers, and the absence of a fuse offers the possibility of connecting transformers of higher power ratings. Thus Brush Switchgear and Brown Boveri have developed ring-main units which utilise circuit breakers without fuse back-up. The Falcon RMU developed by Brush Switchgear utilises SF_6 at a few atmospheres overpressure both as insulation and switch medium, with the transformer tee being protected by an SF_6 circuit breaker. The tripping of the circuit breaker is achieved with an overcurrent device having characteristics similar to a high-voltage fuse. Brown Boveri have developed a CTCV RMU which is similar to the Brush unit, except that the SF_6 used for insulation and switch medium is at a pressure of only 20% above ambient, the transformer-tee-circuit interrupter is vacuum and the tripping is by a 2-phase independent static overcurrent relay (Rondeel and Weissferdt, 1984).

A major factor in the choice of a ring-main unit remains cost. Antagonists of the above developments claim that, in the case of the SF_6 RMU with more sophisticated fuse accommodation, one pays a relatively high price for fuse accommodation and less for interrupting performance, whilst with the circuit-breaker-alone option a high price is paid for interrupting performance, which statistically is only rarely needed. An alternative more economic possibility based upon the I_K-interrupter principle (ASEA, Sweden, and US-based compa-

---

* Production of such units has, however, been limited owing to their being superseded by other developments (Oakes, 1987, private communication)

Impact of SF_6 technology upon distribution and utility switchgear

nies) has been proposed by Rondeel and Weissferdt (1984). This utilises an SF_6 ring-main unit with three load-breaking switches (without fuses), an overcurrent relay and a fuse link intended for short-circuit protection only. The latter could be accommodated, as in the I_K interrupter, within the transformer tank itself, or anywhere between the SF_6 unit and the high-voltage transformer winding. A typical protection scheme is given by Rondeel and Weissferdt (1984), the main advantages of which are claimed to be improved overcurrent protection compared to fuses, and small faults cleared by the switch rather than fuse. With the fuse accommodated in the transformer bushing or tank, cost savings of about 5% are claimed compared with conventionally fused SF_6 ring-main units and 25% compared with circuit-breaker-alone ring-main units.

Fuse–interrupter combinations also occur in power systems for rural distribution. Such systems utilise, for economic reasons, a radial overhead-line network, which is susceptible to interruption of supply due to the effects of lightning, abnormal weather, birds, animals and wind-blown foliage. Protection on such a system is provided by a feeder circuit breaker, local automatically reclosing circuit breakers and fuses (Fig. 7.4c). The automatically reclosing circuit breakers (or auto-reclosers) are required to reclose at least twice following fault detections, and also to have a delayed trip whereby the unit would remain closed long enough to cause fuses further down the line to blow and isolate the faulted section. In this manner, improved continuity of supply can be realised (e.g. Stewart, 1984) for a larger number of consumers. The auto-reclosers are pole-mounted and spring-operated, the spring mechanisms being charged by high-voltage solenoids which can draw their power from the overhead line itself.

SF_6 rotary-arc interrupting units (Fig. 7.6b) are particularly attractive for pole-mounted auto-recloser applications, being economically superior to vacuum-interrupter units. Unlike puffer-interrupter units, the rotary-arc interrupter does not produce back forces on the operating mechanism, which is obviously a distinct advantage for pole-mounted installations. Furthermore, the ability of the rotary-arc interrupter to make numerous makes and breaks without service (~ 2000 at $2\,kA$, 200 at $6\,kA$) is important for auto-reclose duties. In combination with microporcessor monitoring and control, the SF_6 rotary-arc unit has resulted in a mechanically simple unit which is cost effective and versatile in application (on account of the operating sequence and characteristics being easily set or changed in the field).

SF_6 auto-reclosers are available from Brush Switchgear ($12\,kV/13\cdot1\,kA$, $27\,kV/10\,kA$, both at $560\,A$) and NEI (Reyrolle) ($14\cdot4\,kV$). In the Brush unit, the high-voltage solenoid for charging the spring mechanism is enclosed within the SF_6 housing to provide better insulation levels and to protect against climatic influences. The microprocessor for controlling the operating sequence of the recloser is housed in a cabinet mounted seperately from the switch, but electrically connected to it.

7.8 Disconnecting and earthing switches

The function of disconnecting and earthing switches in transmission SF_6 metal-clad systems has already been discussed in Section 6.3.1. The required switching duties (high interrupting current (8–12 kA), relatively low recovery voltage (~ 300 V) for the disconnecting switch; lower interrupting current (1–1·5 kA), higher recovery voltage (50–90 kV) for the earthing switch), combined with numerous current interruptions without loss of insulation quality, make similar demands to those of distribution-system interrupters. Although simple spring-loaded contact separation relying upon free-burning arc interruption (e.g. Ali and Headley, 1984) can meet these requirements and is used commercially, clearly prolonged operation without maintenance relies upon reducing the pre-interruption arc duration. For this reason, Yanabu et al. (1982) and Suzuki et al. (1984) have compared the suitability of various interrupters (free-burning, puffer, self-pressurising, suction and rotary-arc) for disconnecting and earthing duties.

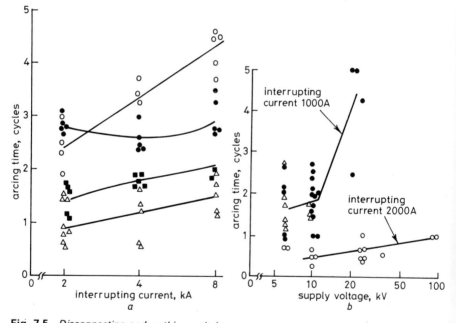

Fig. 7.5 *Disconnecting and earthing switches*
a Comparison of arcing duration of various interrupter types for closed-loop current interruption (Fig. 7, Suzuki et al., 1984)
b Comparison of arcing time as a function of supply voltage of various interrupter types for induced-current breaking tests (Fig. 11, Suzuki et al., 1984)

Test results show that the arcing time and its variation with fault current, for the magnetically driven arc, are shorter than for the free-burning self-pressurising and suction types (Fig. 7.5a). The dielectric strength after several interrup-

Impact of SF_6 technology upon distribution and utility switchgear

tions is preserved, so making the magnetically driven arc interrupter attractive for disconnecting-switch application.

The arcing times as a function of supply voltage are similar for the magnetically driven arc and rapidly opened self-pressurising interrupters, but neither could interrupt for a recovery voltage of about 20 kV. On the other hand, the suction type of interrupter shows a superior performance, and its arcing time does not exceed 1 cycle (Fig. 7.5*b*).

The choice between a simple free-burning interrupter and these more sophisticated alternatives must, of course, be tempered by economic considerations involving an assessment of the improvement which can be obtained in reliability and prolonged life. Furthermore, these considerations for two particular applications illustrate the difficulties associated with the preferential choice of one type of SF_6 interrupter rather than another. It is not clear that, even for a restricted range of applications, each type of interrupter design is optimised or that the designs would need to be optimised differently for each application. Thus, although a considerable amount of information exists upon which optimisation considerations could be based (e.g. flow and magnetic-field geometries, background and induced pressures, current and voltage ranges, contact travel and throttling and gas additives), there still remains a need for additional knowledge before a comprehensive design-choice package can be assembled.

Chapter 8
Operating mechanisms for SF$_6$ circuit-breakers

A summary of the major aspects of the technology of circuit-breaker operating mechanisms is given in Fig. 8.1 in terms of the broad operational requirements, the types of mechanisms available, the system applications and new design aids.

The three most widely used mechanisms are spring, pneumatic and hydraulic. Electromagnetic methods have been investigated, but are not extensively used alone – rather in conjunction with the other techniques. Explosive methods are not popular on account of storage, handling and environmental problems (Eggert et al., 1986).

The four major requirements of a mechanism are low cost, high reliability and reduced maintenance, the ability to be used for a wide variety of duties (e.g. rapid contact separation, close–open cycles etc.) and reduced drive energy. The latter depends in turn upon the kinetic-energy demands of moving mechanical parts, frictional losses and whether the mechanism is used solely to separate contacts or is also expected to provide additional energy for arc extinction, e.g gas compression. This, in turn, depends upon the type of interrupter, the major distinctions being between the 2-pressure interrupter (some gas compression), the puffer interrupter (high gas compression) and self-extinguishing (minimal gas compression). These energy requirements are closely coupled to the type of network for which the interrupter is designed. Thus EHV interrupters, being mainly of the puffer type, are energy expensive, whilst the medium-voltage interrupters, being increasingly of the self-extinction type, require less energy.

8.1 Energy requirements

Development aims are for low-energy mechanisms since both cost and compactness are closely related to the energy requirements. The energy consumptions of the more well established type of circuit-breakers are compared in Fig. 8.2a whilst those for the newer, evolving forms of interrupters are compared in Fig. 8.2b.

Operating mechanisms for SF_6 circuit-breakers

Of the more estabished circuit-breakers, the SF_6 puffer is particularly energy consumptive on account of the need to provide gas-compression energy (Eggert *et al.*, 1986). It requires approximately twice the energy demanded by the other forms of circuit-breakers (minimum oil, air blast and 2-pressure SF_6).

Of the evolving circuit-breaker forms the progressive trend towards lower-energy devices is clearly being realised through the use of the thermal-expansion and rotary-arc type of circuit-breakers (Ruffieux *et al.*, 1986). Thus, whereas the

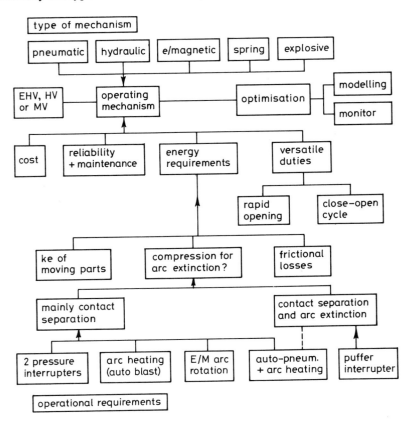

Fig. 8.1 *Technology of circuit-breaker operating characteristics*

auto-pneumatic type of breaker requires typically a kilojoule of energy to ensure proper operation at 40–50 kA fault-current level the combined thermal expansion and rotary arc only requires 300 J for the same fault-current levels. The improvement provided by using either thermal expansion or arc rotation separately is even greater ($\sim$ 70 J compared with 400 J, Fig. 8.2*b*) except that to date the fault current interruptible is limited within the range 12·5–25 kA. This evolutionary process is now well established, but clearly further progress is to be expected in the future.

8.2 Reliability

A CIGRE international inquiry on in-service circuit-breaker failures (Magga et al., 1980) for the period 1974–77 confirmed the commonly held belief that the majority of circuit-breaker failures (70% of major failures and 86% of minor failures) were of mechanical origin, and concerned the operating mechanisms and auxiliary equipment. More recently Eggert et al. (1986) claim that typically 69% of 123 kV circuit-breaker failures are attributable to hydraulic-drive-mechanism faults, decreasing to 48% for 420 kV circuit-breakers (Table 8.1). Of these, 33% were due to oil leakages, 27% to monitor/control-system malfunction, and 8% to pump-motor fault.

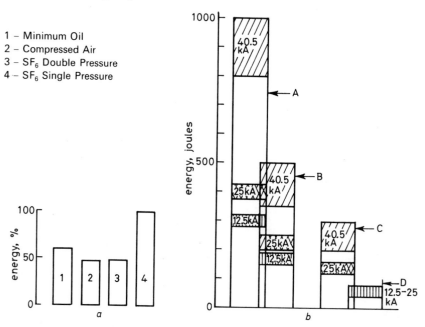

Fig. 8.2 Operating-energy requirements
  a Well established interrupters (Fig. 1, Eggert et al., 1986. Copyright CIGRE)
  b Evolving forms of interrupters (Fig. 13, Ruffieux et al., 1986. Copyright CIGRE)

These statistics are in general agreement with those given by Ikeda et al. (1981), which confirm the predominance of mechanical over electrical failures. Clearly major improvements in operating-mechanism reliability are desirable.

8.3 Puffer-circuit-breaker mechanisms

Although there is an increasing trend to simplify operating mechanisms through the use of arc-heating-induced compression and rotating-arc control, the puffer technique remains important, particularly for the higher-voltage-interruption applications. Since the puffer-type mechanism by its nature probably represents the most demanding and complex technique, it is appropriate to consider the requirements in more detail. The progress made in improving operating mechanisms, coupled with the development of more efficient circuit-breakers, has led to an improvement by a factor of 2 in the interrupting-capacity/operating-energy-of-the-mechanism ratio in the decade up to 1982 (Fig. 8.3).

Table 8.1 Causes of circuit-breaker failure (hydraulic)

Voltage rating (kV)	123	245	420
Interrupter (%)	27	30	28
Drive (%)	69	63	48
Other (%)	4	7	24

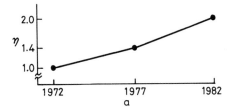

Fig. 8.3 Evolution of interrupting-capacity: operating-energy ratio η (Fig. 7, Leupp, 1983. Copyright CIGRE)

Four basic modes of operation of a puffer-type interrupter may be identified in relation to the mechanical-drive requirements. These correspond to the use of limited nozzle blocking (as in the case of the full-duoflow interrupter of Fig. 3.1c), the use of more severe nozzle blocking (as in the case of the partial-duoflow interrupter of Fig. 3.1b), the use of self-pressurisation (Fig. 3.2b) and the use of suction (Fig. 3.2c).

As already indicated in Section 4.1, a compromise needs to be achieved between flow throttling determined by the moving contact and nozzle size, the pressure generated by the arc and a sufficient puffer volume for low-duty interruption. With all four types of interrupters, long strokes are required to influence gas pressure before contact separation and to maintain it during interruption. Considerable forces are involved – of the order of 20×10^3 N for a current of 50 kA with limited nozzle blocking, and even higher with the more severe blocking (e.g. Fawdrey, 1978). Since the initial gas-compression load with all four types is zero, all the mechanism output is initially available for acceleration of the piston. However, the work input to acceleration is less than half that to gas compression, so that the size of the mechanism is largely determined by the gas-compression load (as already implied by Fig. 8.2b). Hence a flat or rising characteristic is required. For the heavy-duty full-duoflow system, a total energy of typically 5×10^3 Nm (including 15% frictional losses) is required for opening, but considerably less for closing (Fawdrey, 1978). (This is the reverse of the situation with oil circuit-breakers.)

For normal puffer operation it is the piston chamber rather than the piston itself which is moved. In the full-duoflow breaker pre-compression is achieved by using the moving contact as a slide valve, so that the gas flow is not initiated until the contacts separate (Fig. 1.2). Since the arc commutates from the contact fingers to the main arcing ring at an early stage, the contact system has an extremely long life expectancy – of 20 000 interrupter units in service for up to 17 years, no contact needed to be changed (Beier, 1981). In the partial-duoflow breaker pre-compression is achieved by the controlled extracting of one contact through the nozzle, using its movement to dynamically block the nozzle (Fig. 3.1b). Contact separation is usually earlier in the partial-duoflow breaker. The full-duoflow breaker has a longer total travel in order to move the blast cylinder of the contacts.

The relatively long pre-compression phase [$\sim$ 15 ms; Harris (GEC)] before contact separation in the full-duoflow interrupter means that the fault current flows for a prolonged period before circuit interruption can be attempted. In order to reduce this period to achieve interruption within two cycles of the current waveform, the contact travel up to the point of separation may be minimised, whilst to compensate for the accompanying lower compression, the fixed piston may be replaced by a counteracting piston (Beier et al, 1981). This compression is achieved not only by the movement of the piston chamber, but also due to the superimposed action of the moving piston. In this manner the same pressure as the fixed piston alone up to the point of contact separation may be obtained in a shorter time interval. During the arc-quenching phase the operating mechanism ensures that the counteracting piston remains almost stationary whilst towards the end of the interruption phase it is arranged that the piston moves into the off position ready for the next tripping of the circuit-breaker.

EHV systems can only be operated economically if overvoltages are suitably limited during closing (Section 6). To ensure that this condition is satisfied, closing resistors may be fitted in parallel to the break and are set to cut in a few milliseconds prior to the main contacts closing. The system for achieving this sequence of events is mechnically coupled to the drive gear of the main interrupter (Beier et al, 1981). Thus, during the initial part of the closing operation, the current flowing is limited by the resistance. The opening spring of the resistor interrupter is charged at the same time, so that after a preset time of 8–12 ms, the main interrupter closes and short-circuits the resistance. Subsequently, the resistor is decoupled from the drive and opened by its spring. The sequence is repeated following the next tripping of the circuit-breaker. This system ensures that the contacts operate in the correct sequence, and that the resistor is closed for an accurately defined period.

An experimental 1 cycle puffer breaker (245 kV/40 kA) has been investigated

Operating mechanisms for SF_6 circuit-breakers

by Natsui et al. (1980) and shown to be capable of minimum and maximum interrupting time of 0·5 and 1·3 cycles for both 90% short-line fault and 100% breaker-terminal-fault conditions, including symmetrical and asymmetrical current waveshapes. The benefits of 1 cycle operation include a reduction of stresses on major substation equipment, prevention of transformer explosions, reduction of damage due to flashovers and prolonged maintenance interval of the circuit-breaker due to reduced contact erosion. The 1 cycle operation has been achieved through the use of higher SF_6 pressures (15 bar), with a subsidiary suction chamber and a new oil hydraulic operating mechanism using a pilot valve driven by a force motor. Suction is generated immediately after the moving contact commences to travel, but the suction chamber is later exposed in order to prevent over-pressurisation due to the arc-heated gas (Section 4.2.1). Although the performance of the breaker has been convincingly demonstrated, it is unfortunately susceptible to loss of dielectric performance owing to the SF_6 liquefying if the ambient temperature decreases only to 4°C.

8.4 Choice of drive type for puffer interrupters

The choice of particular drive type (i.e. pneumtic, hydraulic, spring) for EHV applications appears hitherto to have been based upon manufacturer prejudice. However, it is claimed that there currently exists a trend by 'leading' manufacturers towards hydraulic drive (Eggert et al., 1986). When the single-pressure SF_6 breaker was first introduced, only 20% of circuit-breakers utilised hydraulic drives, 20% used spring energy and 60% were pneumatically driven. By 1983 the proportion of circuit-breakers using hydraulic drive had increased to 80% at the expense of both pneumatic and spring drives (each 10%).

Spring systems have a characteristic whereby the force decreases linearly from the charged position, which is not ideal for a puffer interrupter unless appropriately corrected through the linkage. A more serious disadvantage is the need for a large tripping force. However, spring systems offer secure storage of mechanical energy, accurate and consistent movement, and obviate the need for valves, gaskets and leakage precautions. It is not yet clear to what extent disadvantages concerning too limited amounts of stored spring energy may dominate to restrict the use of springs to lower current ratings. Nonetheless, for medium-voltage applications SF_6 circuit-breakers in most cases utilise spring drive. Recently Stephanides et al. (1986) claimed that improved drive modelling has led to a reduction of 20% in the spring energy required, leading to a suitable drive for a 63 kA 420 kV SF_6 puffer breaker.

The basic characteristic of a pneumatic mechanism matches the puffer requirement, and is readily modified by dimensioning the air flow. However, experience with air-blast circuit-breakers suggests that problems are likely with moisture and compressors (Fawdrey, 1978). Furthermore, pneumatic mechanisms are noisy and relatively expensive.

140 Operating mechanisms for SF_6 circuit-breakers

The output characteristic of a high-pressure oil actuator is also matched to puffer requirements. Furthermore oil is more readily controlled, except for the initial response. Components of a hydraulic system are much smaller than their air equivalents, and hydraulic mechanisms have a good reputation for long and troublefree service. The most difficult problem to overcome in using high-pressure oil is the pressure transients resulting from rapid valve operation which would be required by 2 cycle circuit interruption. The design problem is then largely concerned with achieving high operating speeds, whilst at the same time minimising stress.

8.5 Modelling puffer-drive mechanisms

Recently there have been several attempts to model theoretically puffer-drive mechanisms, and evidence suggests that practical benefits are beginning to acrue.

From a knowledge of the drive mechanics Kurimoto (1985) has modelled hydraulic and spring-drive mechanisms. Both hydraulic transients were analysed by the method of characteristics, whilst the linkage system was analysed using a stiffness method. More than 80 design parameters were taken into account, including parameters associated with the actuator accumulator, hydraulic valve, buffer, hydraulic pipes, properties of the hydraulic fluid, bell cranks, insulated drive rod and interrupter support column. Typical predictions for the actuator cylinder, hydraulic-drive force and actuator-travel characteristics for a 420 kV/60 kA metal clad circuit breaker are shown in Fig. 8.4a. By combining such mechanical-drive modelling with the predicitons of a simplified empirical model of the arc behaviour, actuator-travel and transient-recovery characteristics as a function of time have been derived. Such calculations have demonstrated that premature deceleration of the drive can occur when used for 3- rather than single-phase operation (Fig. 8.4b). As a result, piston-chamber pressurisation is less under 3-phase compared with single-phase operation, so that the thermal-recovery performance is reduced for longer arcing durations (Fig. 8.4b). As a consequence, the maximum arcing time for successful interruption is reduced from 19 ms to 12·3 ms for the particular case considered.

In the case of a single phase of a puffer breaker, the additional compression caused by nozzle blocking and arc-induced heating can cause the acceleration of the piston to be reduced, and possibly temporarily reversed during peak current arcing. Such effects can be predicted using computer simulations, and show good agreement with the measured behaviour (China, 1984).

Design calculations have been made by Takahasi et al. (1974) concerning magnetic-drive assistance operating a puffer circuit-breaker. The mechanism involves electromagnetic repulsion between a secondary short-circuit ring, carrying the piston and contact, and a primary coil, so that the coupled-circuit method of calculation used for rotary-arc-interrupter field calculations (Turner

Operating mechanisms for SF_6 circuit-breakers 141

and Chen, 1985) may be used to optimise the design of the system. The calculations show that a fairly flat force as a function of displacement may be obtained with a long primary coil and a variable gap between the coil and the secondary short-circuit. The driving force is greatest for high-conductivity materials such as copper or aluminium (Takahashi et al., 1974). The use of electromagnetic-

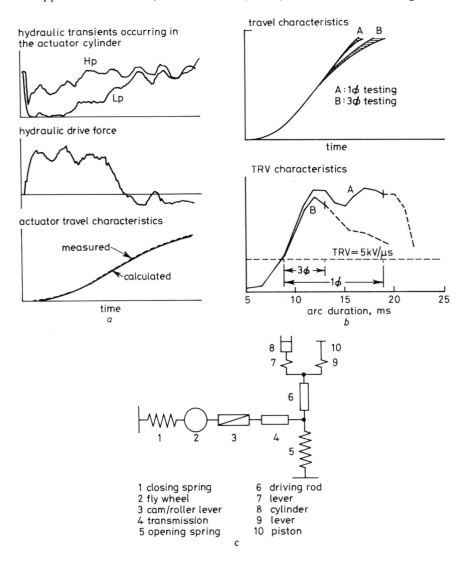

Fig. 8.4 *Modelling of operating mechanisms*
  a *Characteristics of hydraulic mechanism (Fig. 2, Kurimoto, 1985)*
  b *Effect of 3-phase operation on mechanical characteristics (Fig. 4, Kurimoto, 1985)*
  c *Schematic of spring-operated mechanism (Fig. 3, Stephanides et al., 1986. Copyright CIGRE)*

drive assistance can reduce the external force typically by a one-half to two-thirds of that for a conventional puffer breaker, at the expense of a more complicated piston-chamber structure and shorter-duration compression.

The transient behaviour during the closing cycle of a spring-drive mechanism has been modelled by Stephanides *et al.* (1986) with the aim of optimising the energy requirements. During the closing operation with such mechanisms the circuit-breaker contacts are closed and the stored energy of the closing spring is transferred into the opening spring via a flywheel [Fig. 8.4c (Fig. 3, Stephanides)]. Measurement of forces at different locations of the mechanism showed that the occurrence of strong oscillations produced unacceptably high friction, unadvisable stresses, and hence low efficiency of energy transmission. Two separate computer packages were utilised to model the system – the first was used to design the cam (3, Fig. 8.4c) and the second to analyse the transient forces and torques of the whole system. This second calculation scheme was based upon a program for calculating transients in an electrical system based on waveguide theory and using an analogue technique of the type used by Leclerc *et al.* (1980) to relate the mechanical and electrical parameters. As a result, the unacceptable oscillations in the mechanism were reduced, leading to a 20% reduction in the closing-spring energy requirements and an extension of usage of the mechanism to drive a 63 kA 420 kV puffer breaker. Furthermore, the calculations indicate that the physical limitations of such spring mechanisms have not yet been reached, so that the possibility of adopting such mechanisms for even higher ratings remains to be explored.

Chapter 9
Impact of SF_6 technology upon specifications, testing and instrumentation

The development and commissioning of SF_6 circuit breakers and gas-insulated components involves testing electrically, dielectrically and mechanically to meet standard specifications as stipulated by existing standards (IEC, ANSI) and also any additional requirements of the consumer. The advent of SF_6 circuit breakers with the closer coupling between arcing and mechanical drive (Chapter 8) has led to additional test requirements and the introduction of more sophisticated instrumentation.

9.1. Circuit-breaker testing

9.1.1. Electrical tests
Existing standards relating to network-imposed conditions cover terminal-fault interruption, (including inherent transient-recovery voltage), Short-line-fault interruption (SLF) (i.e. fault on an overhead line a few kilometres from the circuit breaker) out-of-phase switching and switching of capacitive load without restrikes (e.g. Table 9.1; Nakanishi *et al.*, 1982). Some actual requirements, which are not yet covered by standards, include evolving fault interruption, parallel interruption of short-circuit currents, switching of small inductive currents without overvoltage generation, and switching of back-to-back capacitors (e.g. Berneryd, 1980).

The three variables, which mainly designate the electrical stress of the circuit breaker, are the interrupted current I, the highest instantaneous voltage after current interruption V, and the time variation of the recovery voltage dV/dt. With a short-line fault, the dV/dt immediately after current zero is particularly severe. With terminal faults, high instantaneous voltages occur during the dielectric recovery phase, and may cause breakdown. The voltage up to breakdown has no effect. Switching low inductive and capacitive currents can produce a high-voltage load during the final phase of the dielectric-recovery period. The influence of the current-flow period is no longer noticeable, and recovery is mainly determined by the increasing contact clearance. Overvoltages may be

Table 9.1 *Phase/tank voltage stresses imposed on tank-type circuit breakers (Nakanishi et al., 1982, CIGRE)*

Switching duty	System condition	Current to be interrupted[1]	Voltage peak to ground[2]
100% BTF		100%	$E \times 1\cdot3 \times 1\cdot4$
SLF		60 ~ 90%	Ex $(1\cdot0 + 0\cdot9 \times 0.4)$ $= 1\cdot36\,E$ Ex $(1\cdot0 \times 0\cdot6 \times 0\cdot4)$ $= 1.24\,E$
Out-of-phase		25 ~ 50%	$E \times 1\cdot25 = 1.25\,E$
Small capacitive current		less than 1%	$1\cdot0E$
Small inductive current		$\simeq 1.0\%$	$1\cdot0E$

[1] Percentage based on rated interrupting current
[2] E = rated voltage $\times \sqrt{2}/\sqrt{3}$

caused by restrikes when switching capacitive currents, and by pre-zero current chopping when switching low inductive currents (Section 7.2).

Dielectric requirements are also well covered by standards, after the introduction of switching-surge tests and bias tests, but the case of voltage across open-circuit breakers for extended periods of time remains to be covered. Some kind of long-term dielectric tests are expected for circuit breakers that are stressed more or less continuously. Mechanical reliability and endurance also need to be assessed.

9.1.1.1 test circuits

Circuit-breaker testing requires high short-circuit power in conjunction with both current and voltage which are variable over a wide range. Direct testing of many circuit breakers is no longer possible on account of the high power, current and voltage levels required. Instead, synthetic testing is used whereby the high currents and voltages, which appear successively at the circuit breaker, are derived from two separate sources connected across the circuit breaker. The correct simulation of the overall test conditions requires numerous accessory devices (auxiliary breakers, trigger gaps etc.), protective equipment and control systems, which need to be accurate to within milliseconds or microseconds.

Impact of SF_6 technology 145

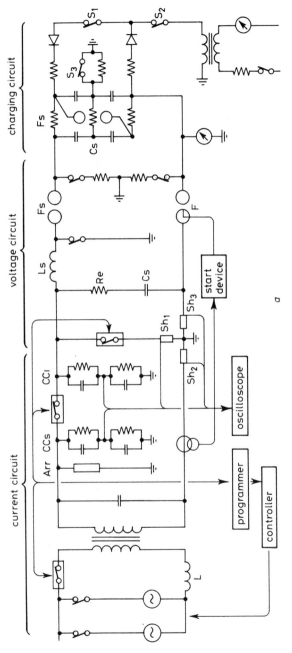

Fig. 9.1 Basic and modified Wiel-type circuits
a Circuit layout to reduce corona loss (Fig. 1, Murano et al., 1974. Copyright IEEE)

146 Impact of SF_6 technology

Even with the synthetic test circuits of modern short-circuit testing stations it is no longer possible to test fully to the highest developed breaking capacities. However, since very high-capacity breakers normally consist of several breaks in series, a single break (with its proportionately lower breaking capacity) can be tested under the partial load which it suffers in actual service, provided that

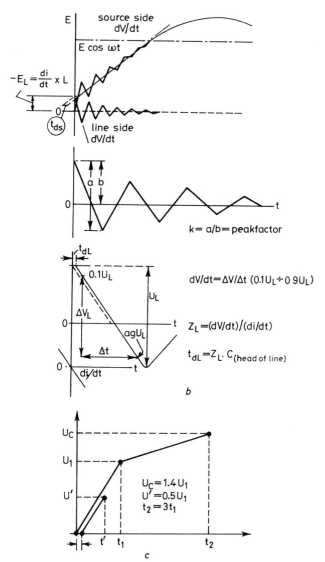

Fig. 9.1 *cont.*

b Details of transient-recovery-voltage waveform (Fig. 1, 3, Van der Linden and Sluis, 1983. Copyright IEEE)

c 4-parameter representation (Fig. 1, Yamashita et al., 1978. Copyright IEEE)

Impact of SF_6 technology 147

the interaction of the multiple breaks during the interruption process is negligible (e.g. Hermann and Ragaller, 1979).

9.1.1.2 Recent modifications to the basic Weil circuit
The most popular synthetic test systems are based upon various forms of the Weil circuit. Modifications are available to limit charging voltage to only one-quarter of the required source voltage so that corona losses during charging are reduced (Murano *et al.*, 1974; Fig. 9.1*a*). Multiple-loop arcing, as required for many newer forms of SF_6 interrupters, may be achieved, despite the limited current-source voltage, by injecting a reverse-current pulse about 100 μs before current zero to enable current commutation (e.g. Murano *et al.*, 1974).

The most severe transient recovery voltage which modern circuit breakers need to withstand is associated with short-line faults. The voltage waveform consists of two superimposed frequencies – a high-frequency sawtoothed component due to the line-side reaction to interruption, and a low-frequency component due to the generator side (e.g. Fig. 9.1*b*). IEC and ANSI regulations stipulate requirements with regard to the parameters of the waveform, e.g. peak factor k (Fig. 9.1*b*) and the form of the initial voltage transient. Whereas ANSI

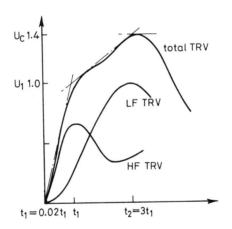

Fig. 9.1 *cont.*
d Low- and high-frequency components required by test duty 4 (IEC) (St. Jean in Yamashita, 1978. Copyright IEEE)

regulations are based upon an EXP–COS waveform (e.g. Flurscheim, 1982), IEC regulations use a 4-parameter representation as shown on Fig. 9.1*c* (Yamashita *et al.*, 1978). Of particular importance is the time delay to restrike, t_d (Fig. 9.1*b* and *c*).

Circuit-breaker testing requires that voltage waveforms to meet these specifications should be available. This is usually achieved by a suitable choice of methods to control the voltage transient produced by the voltage source of the Weil-type synthetic circuit. Several methods have been proposed which may be classified either according to the form of the wave-shaping circuit (i.e. series, parallel or combined series–parallel combination of circuit elements), or according to the methods by which the wave-shaping circuit is introduced. With the latter classification the options are: to perform two separate tests using different 2-parameter waveforms; to connect an additional circuit while the transient recovery voltage is being generated; or to naturally generate the waveform by passive elements connected to a Weil circuit (inherent transient-recovery-voltage method). With series-connected circuits, the current injected from the voltage circuit may have a dual-frequency component, so that the di/dt before current zero may be distorted. The use of artificial lines consisting of π, T or series LC networks has the disadvantage of requiring a large number of elements in order to produce an acceptable triangular waveform (Fig. 9.1b). A line consisting of a limited number of π or LC series elements also has an inherent time delay due to the effective capacitance across the line terminals (Van der Linden and Van der Sluis, 1983). If, alternatively, the method of connecting additional circuit elements during restrike-voltage generation is used, then precisely controlled switching is necessary.

In order to overcome these problems, Yamashita *et al.* (1978) have suggested using parallel capacitors in conjunction with the inherent-transient-recovery-voltage method; i.e. by using an artificial line in series with the normal transient-recovery-voltage control branch of the Weil circuit (Yamashita, 1978). The high-frequency component of the restrike waveform is governed by the artificial line, whilst the low-frequency component (Fig. 9.1d) is governed by the RC combination. The time delay of the restrike waveform t_d may therefore be controlled in isolation from the remainder of the voltage waveform by the resistance connected across the artificial-line terminals (St-Jean in Yamashita *et al.*, 1978).

Van der Linden and Van der Sluis (1983) have developed the artificial line to produce acceptable test-station waveforms. This line is inserted in series between the test breaker and restrike-voltage source and possesses only a low stray capacitance on account of the small size of the network elements, which results in a time delay t_d shorter than $0 \cdot 05\, \mu s$ for a line surge impedance of $450\,\Omega$. A peak factor of $1 \cdot 8$ can be achieved, which meets the specification for low-voltage injection of a 420 kV 80 kA 50 Hz SLF 90%. For unit testing, the surge impedance of the artificial line needs to be proportionally reduced, leading to such low values of the inductance that stray inductance can introduce a ripple on the initial waveform, which gives an unacceptably high initial rate of rise. This may be eliminated by an RC series network between the lineside terminal of the test breaker and ground (Van der Linden and Van der Sluis, 1983).

However, during short-circuit fault interruption in the field, the transient-recovery voltage may contain such high-frequency oscillations of small amplitude due to reflections from the first discontinuity along the busbar (Nakanishi *et al.*, 1982). This produces a more severe initial rate of rise of voltage, which is proportional to the surge impedance and the current. The peak value is proportional to the inductance and the current. Such initial transient-recovery-voltage waveforms may be generated either by increasing the surge impedance of the circuit-breaker-side section of the short-line fault circuit, by using the power cable as the voltage-waveform-generating circuit or by using an additional circuit to the short-line fault circuit Nakanishi et al., 1982). In practice, the energy associated with such oscillations is small, so that the arc resistance is often sufficient to cause strong attenuation.

These basic synthetic test methods have been applied to a number of different circuit-breaker configurations which have evolved with the growth of SF_6-interrupter technology. These configurations include the 3-phase-in-one-tank arrangements used for voltages up to 145 kV, the multibreaks per pole enclosed in a single tank and tank-enclosed disconnecting switches (Section 6.3). Additional tests have also been developed to verify the insulation security of such tank-enclosed systems.

9.1.1.3 Test methods for three phases in the tank breakers

With circuit-breaker designs involving three phases in one tank, it is necessary to verify the recovery characteristics between phases as well as between contacts, since both may be affected by arc-heated gas exhausts. The most severe recovery conditions for such 3-phase circuit breakers occurs under breaker-terminal fault interruption. Under such conditions the arc current, the arc-injected energy and the recovery voltage between phases are all greater than for other duties such as short-line fault and out-of-phase interruptions. Table 9.2 shows the recovery-voltage stresses after a 3-phase short-circuit interruption in a non-effective grounding system. The voltages between the two contacts in an interrupting unit, and between the phase and tank, have a maximum value of 1·5, which occurs after the first pole is interrupted. However, the phase-to-phase voltage has a greater maximum value of $\sqrt{3}E$, which occurs after all three poles have interrupted.

A number of synthetic test circuits have been proposed to simulate these fault conditions. With all such circuits, appropriate arc durations may be obtained with multiloop reignition circuits (Section 9.1.1.2) in each phase. Nakanishi et al. (1982) have used a 3-phase synthetic circuit for stressing a single phase with a 4-parameter transient recovery voltage supplied by a Wiel-type circuit. Alternatively, a 3-phase synthetic circuit may be used in which two of the three phases are connected in series (Nakanishi et al., 1982). Only two multiloop reignition circuits are required, and the third phase of the test breaker is used as an auxiliary breaker. Since the transient recovery voltage is applied to both one terminal of the second phase and the opposite terminal of the third phase, the phase-to-phase voltage can be applied to both moving and fixed contacts. The arc-injected energy with such a test circuit is slightly greater than for the previously described system.

The verification of the interruption performance with both these test circuits is based upon the contact–contact, phase–tank and phase–phase recovery vol-

Table 9.2 *Comparison of voltages between non-effectively grounded system and synthetic test*

Item		Recovery voltage ($\hat{E}$: peak value of phase voltage)								
		Between contacts			Phase-to-tank			Phase-to-phase		
Phase		A	B	C	A	B	C	A-B	B-C	C-A
Non-effectively grounded system	first phase cleared	$1{\cdot}5\hat{E}$	0	0	0	$1{\cdot}5\hat{E}$	0	0	$1{\cdot}5\hat{E}$	0
	last phase cleared	$1{\cdot}0\hat{E}$	$1{\cdot}0\hat{E}$	$1{\cdot}0\hat{E}$	$1{\cdot}0\hat{E}$	$1{\cdot}0\hat{E}$	$1{\cdot}0\hat{E}$	$\sqrt{3}\hat{E}$	0	$\sqrt{3}\hat{E}$
Synthetic test	first phase cleared	$1{\cdot}5\hat{E}$	0	0	0	$1{\cdot}5\hat{E}$	0	0	$1{\cdot}5\hat{E}$	0
	last phase cleared	$\sqrt{3}\hat{E}$	$1{\cdot}0\hat{E}$	$1{\cdot}0\hat{E}$	$1{\cdot}0\hat{E}$	$1{\cdot}0\hat{E}$	$1{\cdot}0\hat{E}$	$1{\cdot}0\hat{E}$	$\sqrt{3}\hat{E}$	0

From Table 1, Morita (1985)

tages being identical to each other and for all three phases. As such, the simulation of the non-effectively-grounded system behaviour (Table 9.2) requires, first, testing with the above systems at a voltage of 1·5E, followed by additional tests to verify the inter-phase recovery at $\sqrt{3}E$, using the system proposed by Nakanishi (1982).

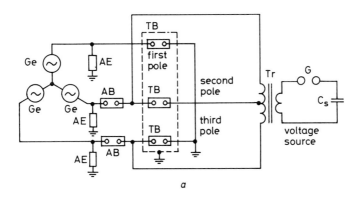

Fig. 9.2 *Test methods for 3-phases in one tank*
a Test circuit using transformer for voltage injection (Fig. 9, Yamamoto et al., 1985. Copyright IEEE)

More recently, Yamamoto *et al.* (1985) (Fig. 9.2*a*) have described a test system utilising low-voltage capacitors stepped up by two transformers applied with opposite polarities, and supplied to both the second and third phases. However, the closest simulation of the non-effectively-grounded system which has been achieved to date appears to be that proposed by Morita *et al.* (1985) and shown in Fig. 9.2*b*). The first phase to clear, A, is stressed by the current injection circuit, V_i. Following clearance of the first phase this voltage decays to a constant value at a rate determined by $C_d R_d$ (Fig. 9.2*b*). This decay is arranged to occur before interruption of the third phase, so that the phase-to-phase voltage does not exceed the required $\sqrt{3}E$ values (Table 9.2). Immediately after clearing the second and third phases, a voltage-injection circuit is connected to both phases with a polarity opposite to V_1 (Fig. 9.2*b*). The voltage between contacts is therefore V_1, the phase–tank voltage (V_1 and V_2) and the phase–phase voltage ($V_1 - V_2$). As a result, the correct dielectric stresses are applied to each part of the circuit breaker (Table 9.2) (apart from the $\sqrt{3}E$ value between the first phase contacts) following clearance of the third phase.

9.1.1.4 Unit testing of multibreak tank-type circuit breakers
Ideally, the performance of a multibreak tank-type circut breaker should be tested by supplying the full current and voltage to the complete multibreak pole of the circuit breaker. Such full-pole testing of high-voltage circuit breakers requires uneconomically large-capacity test facilities. This is in order to avoid current-waveform distortion which may be caused by using a limited current-

source voltage (~ 11 kV) in conjunction with increased arc voltages, produced by the series connection of several interrupting units. The situation is excacbated by test systems utilising series-connected auxiliary breakers which contain additional interrupting units (Yamamoto *et al.*, 1985).

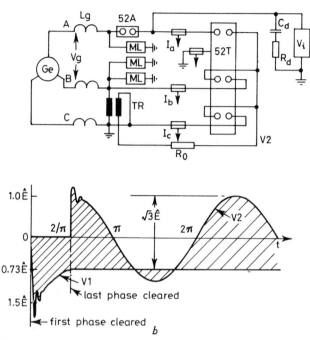

Fig. 9.2 *cont.*
b Test circuit and voltage waveforms for 3-phase fault simulation *(Fig. 1, 2, Morita et al., 1985)*

In order to overcome this difficulty, full pole testing may be substituted by a series of subsidiary tests. These tests should include investigations of the mechanical operating characteristics with full current arcing in each interrupter unit, the voltage distribution across each unit, the interruption performance of a representative unit and the phase-to-tank voltage-withstand characteristics after current interruption. Various synthetic test methods, which have been used or considered for such testing of multibreak-tank-type curcuit breakers, are shown schematically in Table 9.3 (Nakanishi, 1982). Thus the full-pole-interruption test (no.I) may be replaced by combining a unit test (nos. IV, III) with some additional tests.

The close coupling between operating mechanism and arcing conditions in the puffer interrupter, which is most often used in SF_6 tank circuit breakers, requires that it be demonstrated that the mechanism is sufficiently strong to act against the puffer pressure. If tests III and IV (Table 9.3), which involve arcing only at a single pole, are used, then allowance needs to be made for the influence of reduced arc energy through the operation of a single interrupting unit.

Table 9.3 Test circuits for multi-break tank-type circuit breakers

No.	I	II	III	IV	V	VI
Circuit diagrams[1]						
Required capability of current-source power[2]	100	100	50	50	50	50
Required capability of voltage-source voltage[2]	100	50/−50	50/−50	50	50	50/100
Validity of stress on voltage across break[3]	○	○	○	○	○	○
Validity of stress on voltage to ground[3]	○	○	○	×	×	○
Validity of stress on operating mechanism[3]	○	○	△	△	△	○

[1] Schematically shown in case of 2-break-per-pole breakers
I: Current source, V_n: Voltage source, ⊡: Auxiliary breaker
[2] Percentage based on circuit no. I
[3] ○ Valid, △ Valid with reservation, × Not valid
Table 2, Nakanishi et al. (1982. Copyright CIGRE)

154 Impact of SF_6 technology

The distribution of voltage across each interrupter unit has been measured using a potential divider composed of a series of connected resistor–photodiode pairs to provide an optically monitored output (Fig. 9.3). Such tests show that the transient-voltage distribution, is equal to the static distribution which is governed by the grading capacitors etc.

Test performed with the-pole curcuit (I, Table 9.3) and the unit interrupting tests (III, IV, Table 9.3) show that the latter give more severe interrupting conditions and are therefore acceptable as verification tests (e.g. Nakanishi *et al.*, 1982).

The current to be interrupted and the peak phase voltage to ground produced by various switching duties (breaker-terminal fault, short-line fault, out-of-phase switching, small capacitive current, small inductive current) are compared in Table 9.1 (Nakanishi, 1982). This shows that both the interrupting current and phase-to-phase voltage are greatest for the breaker-terminal fault. The phase-to-tank voltage-withstand capability may therefore be verified by solely simulating the breaker terminal fault in conjunction with the specified peak voltage at a given time after current interruption. This may be achieved through the use of a DC, AC or impulse-voltage source in circuits I, II, III or VI (Table 9.3).

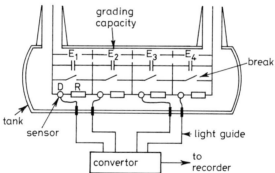

Fig. 9.3 *Measurement across each break of a 4-break tank-type circuit breaker (Fig. 8, Nakanishi et al., 1982. Copyright CIGRE)*

Current distortion with the full-pole test circuit (I) may be reduced by paralleling two generators, in order to increase the current-source transformer output (Yamamoto *et al.*, 1985), so making the test useful for 2-break circuit breakers. For 4-break (or greater) circuit breakers a full voltage system (VI, Table 9.3 with V, replaced by an open circuit and which does not require an auxiliary breaker) is satisfactory (Yamamoto *et al.*, 1985).

9.1.1.5 Synthetic tests for closing and auto-reclosing duties

The purpose of a closing test is to confirm that pre-arcing during the closing operation does not cause contact welding, affect current-carrying capacity etc. Such tests therefore need to be conducted under the rated-voltage condition. A synthetic circuit for closing tests is shown in Fig. 9.4a (Nakanishi, 1982). As the test breaker S_p is being closed, the rated voltage from a voltage source is applied through the impedance Z. When the arc is ignited in the test breaker, the gap S_m is triggered to supply the rated short-circuit current from the current source. The by pass switch S_{H1} is closed to accommodate the more prolonged current flow.

Studies concerning the statistical outage behaviour of a high-voltage system show that most of the short circuits are single-phase-to-ground faults of a temporary nature (e.g. Beehler, 1977; Humphries et al., 1977), which can be cleared by a single-phase opening and reclosing of the circuit breaker without prejudicing the network stability. Although only 10% of the rapid-reclosing operations performed result in unsuccessful reclosing (e.g. Manganaro and Schramm, 1980) nonetheless, the circuit breaker needs to be tested against a short circuit during such unsuccessful auto-reclosing operation. Consequently auto-reclosing tests need to simulate correctly the pre-arcing conditions (e.g. Manganaro and Rovelli et al., 1977; Cazzani et al., 1978) leading to greater flexibility of test circuits than had until recently been demanded. Rigorously, such testing would involve a breaking operation with full voltage and current (designated O_s), followed at some time t later by a making operation C_s and, in the event of non-clearance of a fault, another breaking operation O_{s2} under full voltage and current conditions. However, since the initial breaking operation must in practice be successful, it should be sufficient to simulate that breaking operation using a full current, but reduced voltage, O_D, since it would then only be the full current arcing which would govern the pre-conditioning of the breaker. This enables the test circuit shown in Fig. 9.4b (Managarno and Schramm, 1980) to be used. The circuit is similar to the standard Weil synthetic test circuit with separate current and voltage sources shown on Fig. 9.2.1a, but with an additional loop (high-voltage circuit for C_s) to simulate non-clearance of the fault on auto-reclosing of the test breaker. This loop is energised by the current-circuit generator V_g via a ratio adjuster and step-up transformer (Reg-T), which provides, in conjunction with R, C, the required high voltage across the test breaker prior to completion of reclosure. Once closed, the full fault current is again provided by the current source and, during the subsequent opening operation O_{s2}, full voltage stressing is provided synthetically by the high-voltage circuit for O_s.

Operation of the circuit requires the circuit breakers A_1, A_2 to be closed during the first breaking operation whilst the auxilliary switches A_3, A_4 are open in order to isolate the two high-voltage circuits. Before the making operation, the circuit breaker A_2 must be open and A_3 closed. When the test breaker closes a discharge current flows from the RC branch, and simultaneously the making switch CH is triggered. This switch limits the maximum equivalent test power and influences the minimum and maximum closing angle by means of its time delay and minimum triggered sparkover voltage. AC_sO_s duty can be achieved in 2 cycles or less, a reduction below 1 cycle requiring A_3 and A_4 to operate in less than 5 ms using special switches.

156 Impact of SF_6 technology

The circuit has been used successfully to test a 420 kV SF_6 puffer circuit breaker with two breaks per phase (Manganaro and Schramm, 1980).

9.1.1.6 Short-circuit tests for disconnecting switches
Testing of closed-loop current interruption by a disconnecting switch (Section

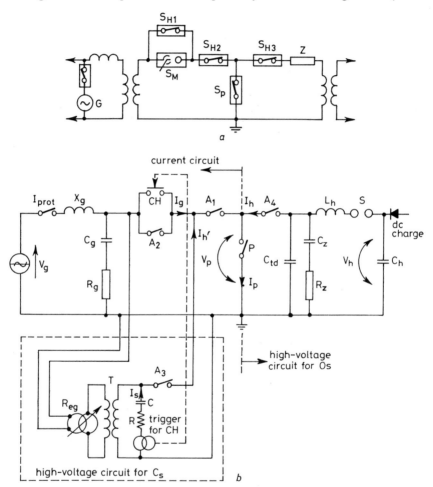

Fig. 9.4 *Test circuits for closing and auto-reclosing duties*
a Closing duty (Fig. 17, Nakanishi et al., 1982. Copyright CIGRE)
b Auto-reclosing duty (Fig. 1, Managarno and Schramm, 1980. Copyright IEEE)

6.3.1) may be achieved with a direct test procedure using the system shown in Fig. 9.5a (Yamamoto *et al.*, 1985), since the voltage levels involved are low (Section 6.3.1). To verify dynamically the insulation to ground, the disconnecting-switch unit is placed on an insulating base and the full voltage applied to the breaker housing from the transformer.

During the slow opening of a disconnector, restriking can lead to a high surge voltage and cause a ground fault to the enclosure (Section 6.3.2). Fig. 9.5b (Yamamoto et al., 1985) shows a circuit for performing a charging-current interruption using lumped capacitance and inductance to simulate a gas-insulated bus.

The maximum restriking surge voltage is theoretically 3 per unit based upon the peak supply voltage (e.g. Yamamoto et a., 1985). This may be achieved by using a sufficiently high value for the bus-simulating capacitance on the load side, C_1, compared with the bushing capacitance, and by using a much higher power-supply-side capacitance C_s than C_1. To simulate conditions for a 500 kV system to meet the 3 per unit surge-voltage requirements is uneconomical using the circuit of Fig. 9.5b.

An alternative test system (Fig. 9.5c) may be used to simulate this most severe condition. This involves fixing the contacts in a predetermined position before charging the load-side capacitance C_1. A charged capacitor SIC then applies an impulse voltage to the power supply side. In this manner, a selection of surge voltages may be applied depending upon the load-side charging voltage or contact gap. Tests on 1100 kV disconnecting switches have been performed (Yamamoto et al., 1985).

More recently Morita et al. (1985) have described a synthetic test method for 3-phase disconnecting switches (Fig. 9.5d). T_1, T_2, T_3 and T_4, T_5 are transfor-

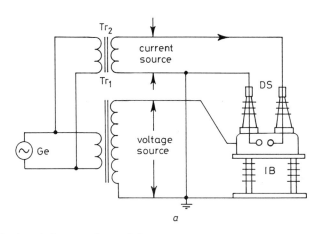

Fig. 9.5 *Testing of disconnecting switches*
a Closed-loop current interruption (Fig. 12, Yamamoto et al., 1985. Copyright IEEE)

mers, T_1 being a tap winding of T_3. The specified 3-phase interrupting current and recovery are produced by T_1, T_2, T_4 and T_5 for the switch DC. The rate-of-rise-of-recovery voltage is controlled by an LC multiple ladder circuit L_z. The high voltage V_e of T_3 is applied during and after interruption as a phase-to-

158   Impact of SF_6 technology

phase voltage. The required and simulated voltages are compared in Table 9.4, showing satisfactory agreement for the contact–contact and phase–phase voltages, but greater phase–ground voltages for the synthetic circuit.

9.1.1.7 Synthetic tests for high-voltage DC circuit breakers
Synthetic methods may also be used for testing high-voltage DC circuit breakers, as well as AC breakers. In the case of AC breakers there is an emphasis upon the equivalence between voltage and current of the synthetic and real systems. With high-voltage DC breakers, the equivalence needs to be extended

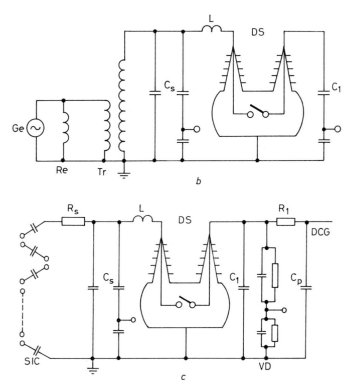

Fig. 9.5 *cont.*
b Charging current interruption (Fig. 14, Yamamoto et al., 1985. Copyright IEEE)
c Simulation of severest condition (Fig. 16, Yamamoto et al., 1985. Copyright IEEE)

to include the energy supplied by the real system, which may be as high as several megajoules (e.g. Yanabu *et al.*, 1981). These conditions can be satisfied by two independent tests, the first concerning the testing of the energy-absorption device (Section 6.4.2) and the second involving the testing of the interrupter unit.

The energy-absorbing device is also required to suppress transient voltages, so that its functions are similar to a lightning-surge arrester (Section 6.3.2). Tests

Table 9.4 *Comparison of 3-phase voltages in direct and synthetic tests for disconnecting switches*

E = phase voltage

Item		Current	Voltage to ground	Phase to phase	between contacts
Direct test	During interrupting	3-phase	E	$\sqrt{3}\,E$	0
	After interrupting	0	E	$\sqrt{3}\,E$	~300 V
Synthetic test	During interrupting	3-phase	$\sqrt{3}\,E$	$\sqrt{3}\,E$	0
	After interrupting	0	$\sqrt{3}\,E$	$\sqrt{3}\,E$	~300 V

Table 2, Morita *et al.* (1985)

160 Impact of SF_6 technology

on such arresters are conducted on a pro-rated unit. This method may be directly applied to the energy-absorbing unit of the high-voltage DC circuit breaker (e.g. Yanabu *et al.*, 1981).

A circuit for synthetically testing the interrupter unit is shown in Fig.9.6a. A low-frequency ($\sim 5\,Hz$) current oscillation is produced by the C_0–DCL combination. The interrupter is open close to current peak to give an arcing time of 10 ms, during which the current variation is acceptably low. The energy-equivalence requirement is therefore satisfied provided that a sufficiently high source voltage is used to avoid current regulation by the arc voltage.

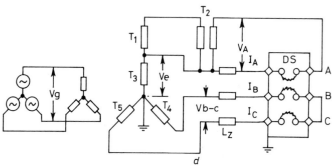

Fig. 9.5 *cont.*
d 3-phase synthetic test (Fig. 6, Morita et al., 1985)

Current interruption is achieved by injecting a high-amplitude current oscillation (Section 6.4.2) from the C, PTr circuit to produce an artificial current zero. After interruption, the transient recovery voltage is produced by the energy stored in the DCL reactor, and is limited by the ZnO arrester. This voltage is maintained across the test breaker and C_1 by operating the auxilliary breaker Ab to interrupt the current flow near the peak recovery voltage. The initial part of the transient-recovery voltage is negative owing to the superimposed current oscillation, before going positive owing to the residual voltage on the commutating capacitor (Yanabu *et al.*, 1981. The transient-recovery voltage, in the case of a hybrid vacuum–SF_6-interrupter unit, is shown in Fig. 9.6b, indicating the manner in which the voltage is shared by each of the interrupters.

9.1.2 Mechanical tests

Mechanical tests are concerned with assessing the performance of the drive mechanism and moving components of the circuit breaker. For instance, with multi-break-per-pole circuit breakers, near simultaneous closure of the gaps is necessary to avoid excessive pre-arcing of late-closing gaps (e.g. Spencer and Harris, 1978), whilst, during opening, each phase should operate together. Specifications require that the opening or closing of each phase should occur within half a cycle of the rated frequency, whilst all contacts of a single pole should open or close within a quarter-cycle of the rated frequency. Although

Impact of SF_6 technology 161

essentially simultaneous operation of contacts is readily achieved by a solid mechanical connection between both breaks and poles, this results in operating times which are too prolonged. Hydraulic and compressed-air connections can ensure practically simultaneous operation in shorter times, but more care is required during manufacture and assembly. Realistic non-simultaneity of 2 ms for the contacts of a pole can be readily achieved to satisfy the above specifications, even after a long period of service. By recording the operating characteristics of the drive mechanism during a number of tests, e.g. initiation of operating signal, start of movement, speed of making or breaking, buffering and completion of movement the diagnosis of a faulty operation is possible.

Since drive mechanisms normally operate at a relatively high speed, they

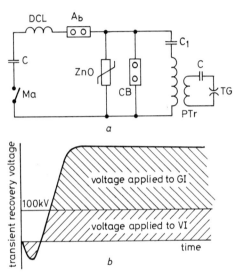

Fig. 9.6 *Synthetic testing of HVDC circuit breakers*
a Test circuit (Fig. 9, Yanabu et al., 1981)
b Sharing of TRV between vacuum and SF_6 interrupters (Fig. 10, Yanabu et al., 1981. Copyright IEEE)

produce excessive vibration and noise, which makes it difficult to detect a mechanical failure during the formative phase. Ikeda *et al.* (1981) have proposed a method to overcome such difficulties, which involves driving the whole of the moving part of the circuit breaker at a very low speed whilst monitoring the driving force, vibratory acceleration, stroke etc. The driving-force measurements provide information about changes in frictional force of the entire mechanism, while the acceleration measurements make it possible to detect any failure of sliding members or abnormal collisions. Some sample test results under normal and abnormal conditions are shown on fig. 9.7d. Under normal conditions, the driving-force characteristic allows the instants of separation of

162 Impact of SF_6 technology

main and subsidiary contacts to be identified (S_1, S_2). With worn contacts, the characteristic is as shown in Fig. 9.7a, whilst play in the linkage mechanism produces the changes shown in Fig. 9.7.

9.1.3 Chemical tests

Monitoring the impurities present in the SF_6 of a gas-insulated system can lead not only to an indication of purity during the comissioning stage, but also to the location of arcing faults during operation. Whereas the comissioning aspects of

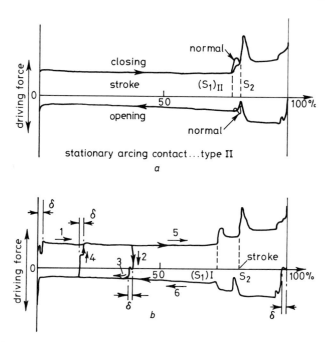

Fig. 9.7 *Mechanical tests*
a Contact wear (Fig. 12, Ikeda et al., 1981. Copyright IEEE)
b Play of pin connections

such tests are concerned with contamination by naturally occurring air, CO_2 and moisture, fault monitoring is more concerned with the decomposition products of SF_6. As a consequence, these latter measurements are also of interest with regard to concern about the toxicity of arced SF_6 (e.g. Sauers *et al.*, 1984).

The recent growth in size and complexity of SF_6 installations has resulted in the need for a significant number of assembly operations to be undertaken on site, so that purity testing during on-site commissioning is essential. On the other hand, the attraction of chemical-purity testing as a means of fault location during operation, as opposed to optical, magnetic, accoustic and electrical methods (Wittle and Houston, 1982), is that detection is possible a long time after fault occurrence, and there is the possibility of differentiating between gas–gap and spacer–surface flashover.

Much work has been undertaken to identify the decomposition products of SF_6 under various electrical discharge conditions (Sauers *et al.*, 1984). This has indicated that all the possible sulphur fluorides (SF_2, S_2F_2, SF_4 and S_2F_{10}) can be produced in an arc discharge, the relative ratios depending upon the nature of, and the power dissipated in, the discharge. The reaction products (or mechanisms) are extremely sensitive to traces of moisture either in the SF_6 gas itself or in the apparatus. The nature of the electrode material influences the initial rate of decomposition of SF_6, whilst the presence of silica, glass, quartz, organic greases, sealing agents, silicone rubbers or any material which can be fluorinated will affect the decomposition products. Several of the decomposition products can be traced to SF_4 as the primary product. For instance, with an electrode material, M, having a valency, n, metallic fluorides, SOF_2, SO_2 and HF are formed in the presence of moisture.

$$\frac{n}{2} SF_6 + M \text{ arc } \frac{n}{2} SF_4 + MF_n$$

$$SF_4 + H_2O \rightarrow SOF_2 + 2HF$$

$$SOF_2 + H_2O \rightarrow SO_2 + 2HF$$

With ceramic insulators containing SiO_2, SiF_4 is produced with H_2O as a catalyst

$$2SF_4 + SiO_2 \rightarrow 2SOF_2 + SiF_4$$

CF_4 may be produced from PTFE nozzles or insulators. SO_2F_2 is only formed in the presence of high moisture contents or on insulator surfaces, but has not been detected in installed gas-insulated systems.

Detection of SF_6 decomposition products in the laboratory may be achieved using such techniques as infrared spectroscopy, mass spectrometry, gas chromatography, nuclear magnetic resonance and electron paramagnetic resonance (Sauers *et al.*, 1984). For circuit breakers and gas-insulated systems, SF_6 decomposition testing is governed through recommendations by IEC and JISC specifications, and CIGRE and the Society of Electrical Co-operative Research, Japan, publications. The four acceptable methods of SF_6 gas analysis include gas chromatography, hydrolysable fluoride, acidity and mineral oil. Gas chromatography involves using helium as a carrier gas with passage through cross-linked polystyrene. Hydrolysable fluoride may be measured spectrophotometrically with alizarin complexon reagent or by the fluoride-ion selective-electrode method (Tominaga *et al.*, 1981). Acidity measurements are made by passage through alkaline solution with a colour indictor reagent, whilst mineral-oil measurements are undertaken by infra-red spectrophotometer of absorption in carbon tetrachloride.

SF_4 may be detected by the hyrolysable fluoride, acidity or chromatographic methods but the latter cannot distinguish between SOF_2. SOF_2 is identifiable by chromatography and detected, but not identified, by hydrolysable fluoride or acidity, whilst SO_2 is detectable by chromatography or acidity. HF may be detected, but not identified, by the hydrolysable fluoride and acidity tests, whilst CF_4, CO_2 and air are detectable only by chromotography.

The amount of decomposed gas depends upon the type of enclosure, the type of metalic electrodes (Fig. 9.8b) and the arcing energy (Tominaga et al, 1981). The concentration of decomposition products is also affected by absorption at the surface of the vessel wall (F^- ion) and by the alumina absorbent which is normally enclosed in each gas-filled section.

However, the SF_6 components are maintained for a sufficiently long period to be detectable by the hydrolysable fluoride and acidity methods. In particular, the inactive gas CF_4 can be used as an index for the past record of the current interruption by the circuit-breaking unit; and earthing, partial discharge or overheating can also be effectively detected (Tominaga et al., 1981). In the case of overheating, experiments performed at temperatures of 250°C indicate that the amount of decomposition products decreases with the type of enclosure material in the order silicon, steel, copper, brass, with aluminium, galvanised and stainless steel showing the lowest levels of decomposition. The use of SO_2 monitoring is promising as a method for differentiating between gas–gap and spacer-flashover faults (Ryan et al., 1985).

The use of SF_6 in electrical equipment produces three types of toxic agents–a single asphyxiating agent (SF_6), a convulsing agent (SO_2F_2) and aggressive pulmonary irritants (S_2F_{10}, SF_4, SOF_2) (Truhaut et al., 1973). The toxicity levels associated with these and other decomposition products may be assessed in of the lethal Dose Low (LD_{LO}) and the Threshold-Limit-Value–Time-Weighted-

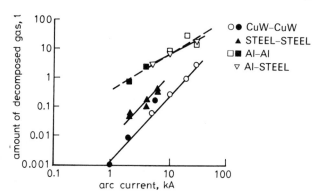

Fig. 9.8 Amount of arc-decomposed gas of various metals (Fig. 3, Tominaga et al., 1981. Copyright IEEE)

Average (TLV–TWA). LD_{LO} is the lowest dose of a substance introduced by any route other than inhalation, over any time period, in one or more divided portions, and reported to have caused death in humans or animals. TLV–TWA is the time weighted average concentration for a normal 8 hour work-day (or 40 hour work-week) to which nearly all workers may be repeatedly exposed without adverse effect.

Confidence in the low toxicity of un-arced SF_6 is gained from its increasing use in diagnostic procedures concerning lung function and respiration (e.g. Jones et al., 1982; Sacknerstal, 1982; Hey and Price, 1982).

The reduction in the toxicity of arced SF_6 by exposure to materials which absorb the lower fluorides of sulphur was demonstrated by Camilli et al. (1952) and confirmed by Boudene et al. (1974). More recently Sauer et al. (1984) have shown that, of the SF_6 by-products, SO_2F_2 and SOF_4 are approximately of the same toxicity and more toxic than SiF_4, which in turn is more toxic than SOF_2. SO_2 and HF did not show toxicity in Sauer et al.'s assay system, although both are known to be toxic to whole animals. Synthetic mixtures composed with concentrations, which were roughly equivalent to, or greater than, that in sparked SF_6, were only weakly cytotoxic, being only about 20% of the cytotoxicity of sparked SF_6. Thus, neither the cytotoxic activity of the synthetic-gas mixtures, nor estimated sums of the cytotoxic activities of the pure individual major components, suffice to account for the cytotoxic potency exhibited by sparked SF_6. This may indicate that highly toxic by-products present in minute quantities may be responsible for the cytotoxicity of sparked SF_6.

Suzuki et al. (1982) have shown that lubricating greases based upon mineral oil are not hardened by the decomposition products of SF_6. However, the oil-retention properties of grease with lithium soap as the thickener decrease with exposure to decomposed SF_6. Grease based upon Bentonite and urea thickeners are not affected.

9.1.4 Particular performance capabilities of SF_6 circuit breakers
Performance investigations using the various synthetic test circuits described above enable a general assessment of the capabilities of SF_6 circuit breakers to be made in relation to the various types of fault conditions (e.g. Table 9.1) and also in relation to other interrupting media (e.g. air and oil). These capabilities, as they existed in 1981, were published by Berneryd (1981), an updated version of which is given in Table 9.5*.

Oil, air and SF_6 circuit breakers all had the capability of interrupting a nominal current of 63 kA. Developments since 1981 have led to SF_6 competing with air blast in being able to interrupt 100 kA, as evidenced by generator-breaker developments (Section 6.4.1). Both air and SF_6 can meet the requirement of a breaking time of 2 cycles (at 50 Hz), and shorter times can be achieved at the expense of complexity in the case of air, and operating energy in the case of SF_6 puffer. SF_6 is better able to meet short-line-fault (Table 9.5) requirements than air-blast, but possibly less so than minimum-oil. Oil and air can cater for initial transient-recovery voltage somewhat better than SF_6, but the latter produces significantly less current chopping than air when interrupting small inductive currents (Table 9.5) and also probably somewhat less than oil. All three media provide adequate capacitive current switching. The high inrush current when a discharged capacitor is connected to a bus plus a capacitor causes no problems for air nor SF_6 but shockwaves in oil may damage the breaker. Media capable of 'soft' interrution of small inductive currents are unlikely to produce an evolving fault.

*[1] Berneryd (1987, private communication).

166 Impact of SF$_6$ technology

Table 9.5 *Comparison of important properties between air-blast, SF$_6$ puffer and oil-minimum breakers for HV and EHV*

Properties	Air-blast	SF$_6$ Puffer	SOV HLR
Interruption			
Breaking capacity	++(100 kA)	+(63 kA)	+(63 kA)
Breaking Time (Cycles)	++($\leqslant$2)	+(2)	−(2.5)
Short-Line Fault	−	+	++
ITRV	+	−	+
Small Inductive Currents	−	+	+
Capacitive Currents	+	+	+
Back-to-Back Capacitors	+	+	−
Evolving Fault	+	+	+
Parallel Breakers	+	+?	+
Current without Zeros	+	−	+
Electrical Endurance	+	++	+
Rated Current	++	++	+(4000 A)
Dielectric			
Insulation to Ground	−	+	++
Insulation Across Open Breaker	− to +	+	+
Mechanical			
Mechanical Reliability	−	+	++
Mechanical Endurance	+	+	+
Erection	−	−	+
Maintenance	−	+	+

Parallel breakers are found in breaker-and-a-half and ring-bus systems (Fig. 6.4a) and it is possible that each may be tripped simultaneously. The fault current would then be shared unevenly and the breaker which initially carried a low current may need to take over the full short-circuit current just prior to current zero. Both air and oil cope well with such a situation, but since SF$_6$ puffer breakers rely upon arcing to assist in gas compression, careful design is required to ensure a proper behaviour. In the case of interruption without current zeros, recent developments have shown SF$_6$ to be at least as satisfactory as oil and air.

Electrical endurance of a circuit breaker is determined by contact wear, erosion of insulation and degradation of the extinguishing medium. SF$_6$ is superior to air and oil in this respect. SF$_6$ and air circuit breakers can carry extremely high continuous currents, since different contacts carry the rated and fault currents. Oil breakers are limited to rated currents of 4 kA.

Concerning dielectric properties, there is no evidence with SF$_6$ of problems with insulation to ground or insulation across an open breaker, provided proper precautions are taken to eliminate water vapour and to ensure no loss of pressure.

Impact of SF_6 technology 167

Since mechanical problems are responsible for most circuit-breaker failures, mechanical reliability is particularly important. Both SF_6 and oil-minimum breakers are superior to air on account of their simpler operating mechanisms. Since mechanical failure rate improves with a reduction in the number of operating parts, SF_6 puffers, with their highest operating voltage per interrupting unit, have a distinct advantage. All three types of circuit breakers (oil, air and SF_6) have acceptable mechanical endurance.

9.2 Instrumentation and diagnostics

The most critical information sought from synthetic short-circuit testing is whether the test breaker has withstood the preset stress and successfully interrupted. This may be determined by measuring the time variation of the stress variables, current and voltage. However, there has been an increasing trend to seek more extensive and detailed information during testing in order to assist in development work. As a result, measurements are made which range from the detailed behaviour of the electrical variables, such as post-arc current and voltage sharing by various units of a multibreak system, to the purely mechanical behaviour of the drive and moving parts, and from the flow system to the behaviour of the plasma column. This trend is having the additional important bonus of bridging the gap between full-power testing information and that obtained in the research laboratories under more ideal conditions for the utilisation of precise but delicate diagnostic techniques.

9.2.1 Electrical measurements

Electrical measurements during full-power testing are normally of overall parameters (e.g. voltage across the breaker contacts or contacts to ground, total current flowing through the circuit breaker), whereas, during the development or research phase, more local measurements (e.g. spatial variation of electric field strength and current density within the arc column) may be made.

In the case of measurements for circuit-breaker-development purposes, total current and voltage values are required over a range of about four orders of magnitude, and, for some conditions, with a time resolution of tens of nanoseconds. Currents are normally measured with resistive shunts. For measuring post-arc currents of a few amperes, a high shunt resistance is advantageous (to give good signal/noise ratio) and a rapid rise time (to follow current change during recognition). Such high-resistance shunts need to be protected from destruction by the flow of current in the several kiloamperes range during the peak current phase. Kobayashi *et al.* (1978) have used a vacuum contactor connected across the shunt and opened immediately before current zero to provide such protection, whilst Stokes (1976) used a low-pressure arc gap and Smith and Jones (1981) a rapid-acting fuse.

With multi-break circuit breakers, a knowledge of the dynamic voltage sharing between the various breaks is clearly of importance during development testing (Section 9.1.1.4). Such measurements have been reported in the past by a number of authors (e.g. Ushio *et al.*, 1972; Calvino *et al.*, 1974) in order to investigate whether the interruption capability of a multi-break circuit breaker was greater than that of a single unit multiplied by the number of breaks. The use of capacitor-type potential dividers has proved popular with many investigators, more recently using series–parallel resistors to measure the voltage- and current-zero points more accurately and with better time resolution (e.g. Kobayashi *et al.*, 1978). Nakanishi *et al.* (1982) have overcome the problem of measuring with respect to a common reference point (which exists with such a potential-divider system) by using series-connected light-emitting diodes and resistors. The signal in this case is optical in nature and is transmitted by optical

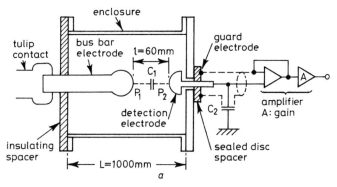

Fig. 9.9 *Electrical measurements*
a Potential sensing unit (Fig. 2, Tokoro et al., *1982. Copyright IEEE)*

fibres, so electrically isolating each element of the voltage divider from the measuring instrumentation. The frequency response of the system is reported as 1 MHz.

Modern instrumentation has also made an impact on on-site circuit-breaker measurements. Tokoro *et al.* (1982) describe a new system used by the Kansai Electric Power Company, which utilises new forms of current and voltage transducers in conjunction with a digital-processing and fibre-optic signal-transmission system. The system is claimed to provide higher-data processing speed, higher-quality system control, ease of expansion, reduced installation space and maintenance. The voltage sensor utilises a capacitor divider of the form shown in Fig. 9.9*a*. This has been used for a 77 kV 3-phase-enclosure SF_6 insulated breaker. Errors arising from capacitance variation due to thermal expansion and mechanical impact on breaker operation are each estimated as 0·1%. Appropriate electrostatic shielding protects against the influence of the other phases. The current-sensing unit utilised inductive coupling to give a 40% reduction in the total volume, compared with a conventional current transformer. For a given accuracy, the dynamic range is an order of magnitude greater than a conventional unit. A 12 bit analogue/digital conversion unit is used to feed the optical-fibre transmission link. Surge arresters are used to protect the electronic system from system overvoltages. The compact nature of the new unit leads to a considerable reduction in the dimensions of a potential transformer unit through the elimination of a bus section and a disconnecting switch (Tokoro *et al.*, 1982).

Murase *et al.* (1985) have developed a method for measuring the form and magnitude of transient voltages induced by disconnecting-switch operation. Such switches generate high-level and high-frequency surges (Fig. 6.5*d*), which, because of the coaxial cylinder geometry of a gas-insulated system, suffer little attenuation during propagation through the system. A number of authors (e.g. Bosotti *et al.*, 1982; Boggs *et al.*, 1982; Yoshizumi *et al.*, 1982) have utilised special electrodes installed within the gas-insulated tank for monitoring such surges. Measurements at frequencies in excess of 500 MHz are claimed by Boggs and Fujimoto (1985). However, for in-service monitoring, such electrodes are

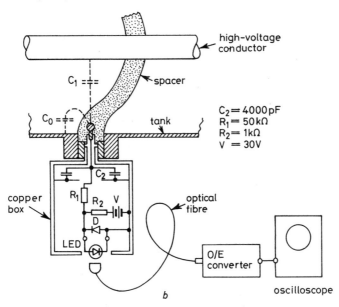

b Surge-measuring equipment (Fig. 3, Murase et al., 1985. Copyright IEEE)

undersirable owing to extra cost and the need for system modifications. This has been overcome by Murase *et al.* (1985) utilising shield electrodes which are already installed and embedded in spaces at various locations within the system. The embedded shield electrodes have a stray capacitance of 2–10 pF to the conductor and several hundred picofarads to the tank, so forming an appropriate capacitor divider. (Fig. 9.9*b*). An additional capacitor C_2 is connected as part of the low-voltage limb of the divider. A resistor R is used to convert the terminal voltage of C_2 into a current which flows through a light-emitting diode electrically biased to respond to voltages of both polarities. Emission from the diode is arranged to be proportional to the C_2 voltage, and is transmitted via an optical fibre to remotely located processing electronics. The bandwidth of the

170 *Impact of SF₆ technology*

measuring system is claimed to be 0·8 kHz to 16 MHz. The system has been used for on-site measurements on a 500 kV gas-insulated system, with estimated transient rates of voltage change $\sim 10^{14}\,\text{Vs}^{-1}$ and response times as high as 5 ns.

A newly developed fault arc-detection system is described by Kopainsky and Muri (1983), which enables the compartment of a gas-insulated system in which the arcing has occurred to be identified. This is achieved with inspection windows fitted to each compartment which carry 50 m-long fibre-optic cables. Light from a fault arc is transmitted through the fibre to the control cubicle for registering. Response to normal switching operations is prevented by an AND gate.

Local electrical-arc properties have mainly been measured in research laboratories, and are intermediate between the overall electrical measurements made in the short-circuit test station and the more fundamental plasma-property measurements of the research laboratory. A knowledge of the axial distribution of voltage along the arc column is important in determining the locality which governs arc quenching (and hence circuit interruption) and at which power-dissipative mechamisms are at a maximum. Such voltage distributions have been measured using voltage probes immersed in the plasma (e.g. Chapman, 1977). Current-density measurements provide an indication of the threshold for the blocking of a nozzle by the electrical cross-section of the plasma arc column. Such measurements have been obtained using small inductive probes placed in the locality of the arc column (e.g. Barrault and Jones, 1974). The electrical cross-section of the arc column at various axial locations and as a function of time has also been measured with a radio-frequency probing technique (e.g. Dhar *et al.,* 1977). Axisymmetric disturbances of the arc column have been detected using such radio-frequency probing, whilst rotation of the arc plasma has been confirmed using inductive probes (Shishkin and Jones, 9185).

9.2.2 Mechanical-drive measurements
A number of methods have been used in the past for monitoring the movement of the mechanical drive of gas-blast circuit breakers; e.g. potentiometric measurements using a link to the slider of the potentiometer. More recently, efforts have been made to measure the contact movement directly, rather than through the associated mechanical linkages.

Spencer and Harris (1978) have developed a digital-recording timer for monitoring such contact movement (Fig. 9.10a). The timer consists of a shutter carrying a number of holes through which an infra-red beam passes at pre-determined dispositions of the shutter. This provides a series of electrical pulses corresponding to each of the characteristics of the mechanism listed above, and whose timing can be electronically compared with those of a correctly operating mechanism.

A miniature travel recorder which can be mounted directly on the contact stalk within the breaker has been developed for monitoring contact travel directly and without interference from linkages (Fig. 9.10b). The unit consists of

Impact of SF_6 technology   171

a coded wheel rotated by the linear movement of the contact stalk (Fig. 9.10b). The rotation is measured optically, the input and output signals being transmitted by optic fibres which ensure insulation integrity. Signal processing is achieved by a microprocessor. A schematic of the system is shown in Fig. 9.10c, along with a sample travel record Fig. 9.10d.

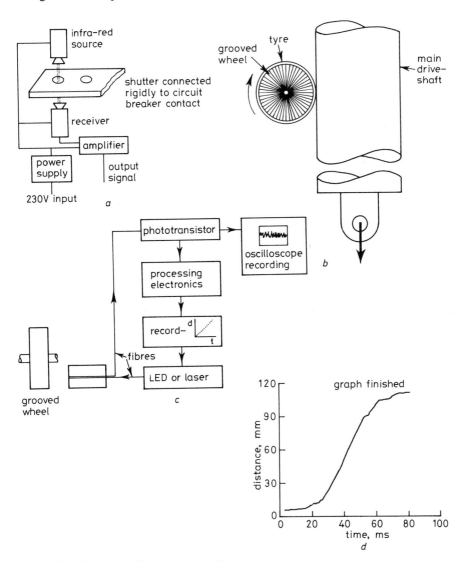

Fig. 9.10 *Mechanical-drive measurements*
  a Optical digital recorder (Fig. 4, Spencer and Harris, 1978)
  b Schematic of optical travel recorder (Shimmin, private communication)
  c Measurement system (Shimmin, private communication)
  d Typical result (Shimmin, private communication)

172 Impact of SF$_6$ technology

9.2.3 Aerodynamic measurements

Aerodynamic measurements are concerned with ensuring adequate flow for arc control and quenching. Pressure measurements are increasingly being used to

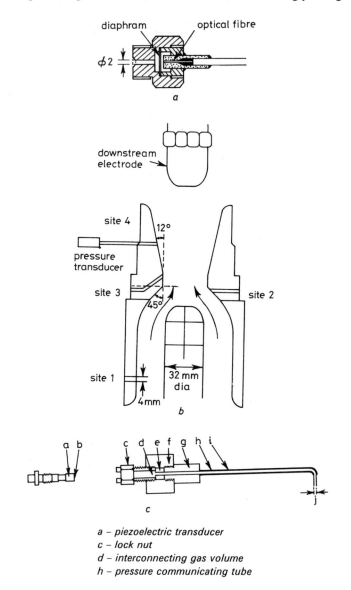

Fig. 9.11 *Aerodynamic measurements*
 a Nozzle pressure using optical transducer (Fig. N-8, N9, Noeske et al., 1983)
 b Pressure-measuring sites on circuit-breaker nozzle (Fig. 4.1, Taylor, Private communication)
 c Pressure probe for arc plasma studies (Jones et al., 1974, Private communication)

a – piezoelectric transducer
c – lock nut
d – interconnecting gas volume
h – pressure communicating tube

monitor flow conditions in puffer circuit breakers. Piezoelectric, piezoresistive and optically based transducers have been used for such measurements (Barrault and Jones, 1974; Jones *et al.*, 1982; Noeske *et al.*, 1983; Taylor 1983). The optical technique is based upon measuring the displacement of a diaphram through the intensity modulation of light transmitted by an optical fibre (Fig. 9.11*a*). Although this technique overcomes the electromagnetic pick-up problems of the more conventional electronic transducers, it suffers from questionable optical-system security particularly in the presence of arc-generated light and from long-term ageing deterioration. The electronic-type pressure transducers need to be electrically and thermally protected from the arc discharge through the use of interconnecting tubes, and possibly filled with vacuum oil. By careful design, such systems can be designed to be virtually distortion-free, provided correction is made for acoustic transit times (Jones, 1984). A system used by Ali *et al.* (1985) during development testing of a 420 kV 63 kA 2-break puffer breaker is shown in Fig. 9.12*b*.

Systems utilising interconnecting tubes have been used, not only for monitoring pressure in the piston chamber, but also at various locations along the main interrupter nozzle (e.g. Taylor, 1983; Noeske *et al.*, 1983;) (Fig. 9.11*b*), within the arc plasma column (Fig. 9.11*c*) and at the upstream contact tip (Fig. 9.11*d*).

Aerodynamic and thermal fields have been mapped in units, suitably modified for optical access, using shadowgraphic, Schlieren or interferometric techniques (e.g. Noeske *et al.*, 1983; Walmsley and Jones, 1980). Such techniques allow the expansion of thermal volumes around the arc, flow fields and shock waves to be identified and related to the instantaneous electrical behaviour. Flow-field visualisation has enabled Ancilewski *et al.* (1984) to identify a more rapid radial inflow at the nozzle inlet than hitherto predicted, and a situation whereby the radial-flow recovery does not match the arc-cross-section collapse. This would appear to be directly reflected in the performance characteristic of the circuit breaker (Section 5.2). More direct measurments of plasma and gas-flow velcoities have been obtained using laser Doppler velocimetry and the oscillatory movement of shock waves in sympathy with the current waveform detected (Todorovic and Jones, 1985). The measurement of turbulence scale and lifetime have been made through the cross-correlation of optical-perturbations propagation along the arc with the imposed flow (Niemeyer and Ragaller, 1973).

9.2.4 Radiation measurements

Analysis of the radiation emitted by the arc discharge provides an attractive means for non-invasive measurements. This analysis may be made at various levels of sophistication ranging from simply the spatial extent of the plasma (using high-speed photography), through the magnitude of the total radiation loss (using thermopiles and flat-wavelength-response photomultipliers) to detailed spectral analysis (using optical spectrum analysers) (Jones, 1988).

High-speed conventional and image-convertor cameras have been increasingly used for measuring arc-column cross-sections during the high-current phase (for nozzle-blocking information) and during the current-zero period for identifying the location of arc rupture (e.g. Hermann et al,1976; Jones, 1984; Lewis et al., 1985). The movement and shape of the arc in a rotary-arc circuit breaker has also been monitored to establish its 3-dimensional behaviour (e.g. Spencer et al., 1985). Such measurements have been mainly limited to the research and development laboratories on account of the need to make design changes to gain optical access to the circuit-breaker enclosure. Fig. 9.12a shows a system of narrow sealed slits in the nozzle wall to gain such access to the nozzle arc. Most efficent use of the film area can be achieved using a number of mirror slices to magnify the slit area, whilst simultaneously reducing the width between slits on the film surface (e.g. Jones, 1984). Transparent nozzles have also been used, but care is needed to calibrate the system to overcome image distortion. Furthermore, the influence of nozzle material (e.g. through ablation) cannot be investigated with such nozzles since realistic nozzle materials (e.g. PTFE, Cu, C) are not sufficiently transparent.

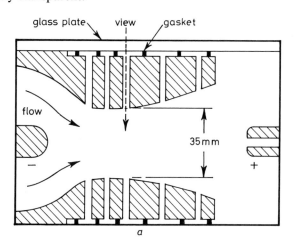

Fig. 9.12 *Radiation measurements*
a Nozzle for arc viewing (Fig. 1 a, Lewis et al., *1985)*

The use of rapid-response photomultipliers in conjunction with such slotted nozzles allows the total radiation reaching the nozzle wall to be determined (e.g. Jones, 1984). Radiation levels as high as 10 MW have been measured, which are sufficient for causing severe nozzle ablation. The use of such methods, with narrow-band optical filters allows the temperature of the ablated nozzle material to be determined (Jones et al., 1986).

The use of spectroscopic techniques can yield information not only about fundamental plasma properties, such as temperature and electron density, but also about contact- and nozzle-material entrainment and vapour jetting. Recent measurements have indicated that, following high peak arc currents ($\sim 53\,\text{kA}$), the temperature, as well as the cross-section, of arc column at current zero is significantly increased and has a direct influence in reducing the interrupting

Impact of SF$_6$ technology 175

capability (Lewis *et al.*, 1985). During the high-current arcing phase under monoflow conditions, a substantial electrode jet occurs having a high concentration of electrode material ($\sim 70\%$) at a temperature somewhat below that of the surrounding plasma (e.g. Jones, 1983).

Until recently the use of such sophisticated spectroscopic techniques was limited to the research laboratory. However, the increased availability of suitable optical fibres has enabled radiation and spectroscopic measurements to be made on commercial circuit breakers undergoing development testing at industrial short-circuit test stations (Ali *et al.*, 1985). Ali *et al.* (1985) utilised a series

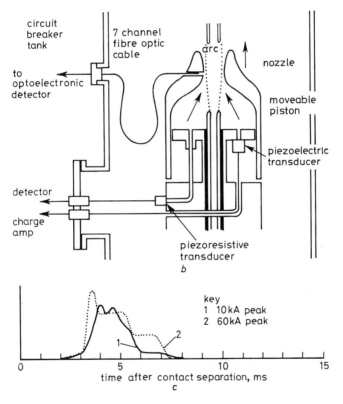

b General arrangement of diagnostics (Fig. 8, Ali et al., 1984. Copyright IEEE)
c Copper-line emission from partial duoflow circuit breaker (Fig. 11, Ali et al., 1984. Copyright IEEE)

of fibres to give optical access to the arc within the nozzle throat (Fig. 9.12*b*) indicating the instant at which the contact passes through the nozzle throat and the increase in intensity with peak current. Spectroscopic measurements at a wavelength of 515.3 nm, which corresponds to a Cu I spectral emission line, shows little increase in emission intensity with peak current (Fig. 9.12*c*), so indicating the effectiveness of the duoflow system in this type of circuit breaker for limiting electrode-vapour entrainment into the intercontact gap.

The use of such techniques under real test-station conditions has strengthended considerably the link between on-site and research-laboratory-type measurements, so that fundamental dscharge information can be more directly transferred to the real interrupter situation.

9.2.5 Chemical measurements

As indicated in Section 9.1.3, the decomposition products of SF_6 may be monitored by gas chromatography and hydrolysed fluoride and acidity measurements. Since these methods are too troublesome to be used for on-site testing, simpler techniques have been proposed which rely upon chemical reactions involving the different decomposed species.

Perry (1981) and Tominaga *et al.* (1981) report the use of a thin-film sensor for detecting the by-products of arcing or partial discharge. The sensor measures the change in resistivity of a $0.6\,cm^2$ wafer of alumina substrate covered by a thin film of polymer. In the presence of pure SF_6, a voltage of 10–50 V produces a current of only 10^{-12}–10^{-14} A, whereas following decomposition due to arcing, the current is increased to 10^{-4} A, depending upon the extent of decomposition. The increased conduction is believed to be caused by higher ionic conduction in the polymer–substrate surface following the absorption of HF (due to the SF_4 and H_2O reaction) and H_2O. Removal of the decomposition products results in a rapid recovery of the sensor. One problem with such a sensor, identified by Tominaga *et al.* (1981), is its susceptibility to long-term instability.

Tominaga *et al.* (1981) have developed a monitor utilising a detecting element composed of alumina powder containing bromocresol-purple as an indicating reagent, whose colour changes from bluish purple to yellow in the presence of contamination. The monitor allows about 10–30 litres of contaminated gas at NTP to flow at a rate of 2–5 litres/min along the length of the detector. The length of the discolouration suffered by the detecting element indicates the degree of decomposition. Elements with different sensitivities are available, so that earthing faults (10 parts in 10^6 weight as HF) and partial discharge activity (0.03 parts in 10^6 weight as HF) can be dstinguished. Such monitors have been used on site with 84–550 kV SF_6-insulated systems.

Ryan *et al.* (1985) have reported tests relying upon the monitoring of SO_2 concentrations using a Gastec Corporation SO_2 detection tube no. 5 La (range 0.5–5 parts in 10^6 with a threshold of 0.1 parts in 10^6). The detection relies upon producing hydrogen chloride with the aid of Barium chloride, according to the reaction

$$SO_2 + Ba\,Cl_2 + H_2O = Ba\,SO_3 + 2HCl$$

The hydrogen chloride is then detected by a colour change from blue to yellow using bromocresol-green indicator (pH values of 5.4 to 3.8 causing the colour change). Although other SF_6 decomposition products could, in principle, affect the colour change, their influence is believed to be minimal. Many will not react with the pH indicator. Others, like the oxyfluorides of sulphur, do not readily hydrolise to form HF, which, anyway, because of its weak acidity and low concentrations, is unlikely to have such a pronounced effect as HCl. Consequently, the proposed detector, with its sensitivity to SO_2, is believed to be capable of differentiating between gas–gap and insulator flashovers, on account of the increased SO_2 production associated with arcing on such insulators (e.g. Boudene, 1974; Gayet, 1984). Baker (1980) reported that arcing in SF_6 gas using aluminium electrodes did not produce sufficient SO_2 for detection, and Ryan *et al.* (1985) have detected an increase by three orders of magnitude in the SO_2 concentration with a bauxite spacer, compared to a gas–gap breakdown.

Chapter 10
Conclusions

The developments in circuit-breaker technology, which have occurred during the last decade as a result of the increased use of SF_6, have resulted in outstanding economic benefits and changes in substation design and appearance which would have been unrecognisable two decades ago. The extension of SF_6 technology to the lower voltage levels is also producing important improvements in system security. If present trends continue, it is likely that SF_6 as an interrupting medium will be substantially available in an economically competitive manner throughout the entire range of switchgear applications listed in Fig. 1.1. The new SF_6 interrupter designs, based upon novel principles such as self-pressurisation and electromagnetic rotation, would appear to be at an early stage of the evolutionary process. Nonetheless, the progress made with such design development, in relation to the research and development efforts, is quite remarkable when compared with the development of, for instance, vacuum interrupters.

The availability of alternative SF_6-interrupter forms (such as the puffer, suction, self-pressurising and rotary-arc types), from which manufacturers can choose, has emphasised the important role which feasibility studies and computer-aided methods can fulfil before more expensive performance testing is profitably undertaken. The practice employed in the past, for economic reasons, by many manufacturers, of 'stretching' existing interrupter concepts to different ratings (e.g. auxiliary venting of electrode vapour through a hollow upstream contact) can, in such an evolutionary climate, lead to over-complex, more expensive and less reliable designs (Noeske et al., 1983). In such circumstances, the path through research and computer-aided-design concepts can produce a lower-cost product (Noeske et al., 1983). There is evidence of increased usage by many manufacturers worldwide of such computer-aided methods in conjunction with modern measurement technology. With the advent of the new technologies, such trends are surely destined to find increased usage for switchgear development in the forseeable future.

Despite the slower uptake of SF_6 technology for distribution applications, there are already indications of important impacts which are likely in the future through the provision of better control to reduce outage times. The advent of

Conclusions 179

SF_6 and vacuum interrupters, with the sealed-for-life concept and ability to undertake a substantial number of breaking operations without degradation of performance, does not prohibit the use of the circuit breaker for switching duties, as has been particularly the case with oil breakers. Since this vital factor in the protection chain is now less of a limitation, it is likely that much effort to produce more efficient and less expensive fault-detection and control systems, based upon new technologies, will be witnessed increasingly in the near future. Indeed, the initial steps in this direction are already apparent in the recent availability of microprocessor-controlled auto-reclosers and in the introduction of optical fibres for monitoring applications.

At transmission-voltage levels the projected increased power transmission demands (Figs. 1.2*a*, *b* and *c*) are leading to ever-increasing transmission voltages, the introduction of high-voltage DC transmission and increased power-system density (and hence potential fault currents). Although such developments may well be limited in number, they do nonetheless appear to be necessary. To cope with such demands, substantial and expensive development programmes are likely to be needed which are approaching the capability limits of even the most advanced synthetic-test facilities. Further developments in this area are therefore likely to be governed by identifying means of overcoming such difficulties through, perhaps, national and international collaboration on a scale hitherto not experienced in the field of circuit-breaking. Since the preparation of this manuscript such major changes in the international switchgear field have indeed occurred with major manufacturers such as the BBC and ASEA combining their activities. Should this prove to be the way forward, it is to be hoped that the UK manufacturers will be major participants in such ventures.

Chapter 11
References

AIREY, D.R., GARDNER, G.E., and URWIN, R.J. (1976): 'Development of SF_6 arc interrupters'. 11th University Power Engineering Conference, Portsmouth, UK
AIREY, D.R., KINSINGER, R.E., RICHARDS, P.H., and SWIFT, J.D. (1976): 'Electrode vapour effects in high current gas blast arcs', *IEEE Trans.*, **PAS-95**, pp. 1-13
ALI, S.M.G., and HEADLEY, P. (1984): 'Developments in modern EHV switchgear', BEAMA – Technical Seminar for PLN (Jakarta)
ALI, S.M.G., RYAN, H.M., LIGHTLE, D., SHIMMIN, D.W., TAYLOR, S., and JONES, G.R. (1984): 'High power short circuit studies on a commercial 420 kV, 60 kA puffer circuit breaker', *IEEE Trans.*, **PAS-204**, pp. 459-467
ALI, S.M.G., HEADLEY, P., LIGHTLE, D., and RYAN, H.M. (1982): 'Arc interruption studies on heating duty SF_6 puffer circuit breakers'. Proc. 7th Int. Conf. on Gas discharges and their applications (London)
ANCILEWSKI, M., BURLIK, A., and KING, L.H.A. (1984): 'Calculation of SF_6 interrupter performance for thermal reignition criteria'. CIGRE Paper 13-07
ANDERSEN, B.R., and SIMMS, J.R. (1983): 'Transient over-voltage protection in metal-clad SF_6 insulated substations', *GEC J. Sci. & Technol.*, **49**, pp. 28-38
ARMSTRONG, A.G.A.M., and BIDDLECOMBE, C.S. (1982): 'The PE2D package for transient eddy current analysis', *IEEE Trans.*, **MAG-18**, pp. 411-15
AZUMI, K., KUWAHARA, H., SAKON, I., MARUTANI, T., and NIWA, H. (1980): 'Design considerations on three phase enclosure type gas insulated substations', *IEEE Trans.*, **PAS-99**, pp. 947-54
BAKER, A., DETHLEFSEN, R., DODDS, J., OSWALT, N., and VOUROS, P. (1980): 'Studies of arc byproducts in gas insulated equipment'. EPRI Report EL-1646
BARRAULT, M.R., and JONES, G.R. (1974): 'Practical arcing environments – arc plasma diagnostics' in VUJNOVIC (Ed): Proc. Invited Lectures 7th Yugoslav Symp. and Summer School on Physics of Ionized Gases (Rovinj, Yugoslavia) (Inst. of Physics, Zagreb Univ.)
BARTLE, J., and JOWETT, B.F. (1984): 'A review of distribution substations in the north east – past, present and future'. Proc. IEE Symp. on Trends in modern switchgear design 3·3-150 kV, Newcastle, pp. 11.11-11.7
BEEHLER, J.E. (1977): 'Seven year summary of EHV power system outages'. CIGRE SC 13 (Pozuan)
BEIER, H., LUHRMANN, H., and MARIN, H. (1981): 'Development of SF_6 circuit-breakers at Siemens', *Siemens Power Enging.*, **3**, Special issue 'high-voltage technology', pp. 29-36
BERNERYD, S. (1981): 'Performance of different types of circuit breakers with respect to network requirements', *J. Elect. & Electron. Enging. Australia*, **1**, pp. 1-7
BLOWER, R.W. (1986): 'Distribution switchgear', Collins, London
BLOWER, R.W. (1984a): *IEEE Trans.*, **PA3-103**, pp. 2753-2761

BLOWER, R.W. (1984b): 'SF$_6$ fuse switches and ring main units'. Proc. IEE Symp. on Trends in modern switchgear design 3·3-150 kV, Newcastle, pp. 15.1-15.5
BLOWER, R.W., CORNICK, K.J., AND REECE, M.P. (1978): 'The use of vacuum switchgear for the control of motors and transformers in industrial systems'. IEE Int. Conf. on Development of distribution switchgear
BRAND, K.P. (1982): 'Dielectric strength, boiling point and toxicity of gases – different aspects of the same basic molecular properties', *IEEE Trans.*, **EI-17**, pp. 451-456
BRAND, K.P., EGLI, W., NIEMYER, L., RAGALLER, K., and SCHADE, E. (1982): 'Dielectric recovery of an axially blown SF$_6$ arc after current zero: Part II Comparison of experiment and theory', *IEEE Trans.*, **PS-10**, pp. 162-172
BRAND, K.P., and KOPAINSKY, J. (1979): 'Breakdown field strength of unitary attaching gases and gas mixtures', *Appl. Phys.*, **18**. 321-333
BRIGGS, A.C. (1985): 'Liquid SF$_6$ injection: A possible solution for the SLF limiting phenomena of the gas blast circuit breaker'. Proc. Int. Conf. on Gas discharges and their applications, Oxford (Leeds University Press)
BRIGGS, A.C., KING, L.H.A. (1979): 'Factors affecting thermal breakdown in gas blast interrupters'. Proc. Int. Conf. on Developments in Design and Performance of EHV Switching Equipment (London, IEE), pp. 52-55
BROWN BOVERI (1986): 'SF$_6$ generator circuit-breakers type HEK/HE1'. Publication CH-A 511 291 E
BOGGS, S.A., CHU, F.Y., and FUJIMOTO, N. (1982): 'Disconnect switch induced transients and trapped charge in gas insulated substations', *IEEE Trans.*, **PAS-101**, p. 3593
BOGGS, S.A., and FUJIMOTO, N. (1985): Disconnect switch induced transients and trapped charge in gas insulated substations', *IEEE Trans.*, **PAS-101**, p. 3593
BOGGS, S.A., and FUJIMOTO, N. (1985): Discussion in Murase *et al.* (1985)
BOGGS, S.A., and FUJIMOTO, N. (1985): Discussion in Murase *et al.* (1985)
BOSOTTI, O. *et al.* (1982): 'Phenomena associated with switching capacitive currents by disconnectors in metal enclosed SF$_6$ insulated switchgear'. CIGRE Paper 13-06
BOUDENE, C., CLUET, J-L., KEIB, G., and WIND, G. (1974): *Rev. Gen. Elect*, June, p. 45
BURGER, U. (1979): 'Insulation co-ordination and selection of surge arrester', *Brown Boveri Rev.*, **4**, pp. 271-280
BROSS, E. (1981): 'High voltage circuit breakers with SF$_6$/N$_2$ gas mixture for extremely low ambient temperatures'. Electrical apapratus section, Switchgear subsection, Canadian Electrical Association, Toronto, Canada
CALVINO, B., MAZZA, G., MAZZOLENI, B., and VILLA, V. (1974): 'Some aspects of the stresses supported by HV circuit breakers clearing a short circuit'. CIGRE Report 13-08
CALVINO, B.J. (1974): 'Study of current zero phenomena under various stress conditions of a single pressure SF$_6$ circuit breaker'. CIGRE Report 13-07
CAMILLI, G., GORDON, G., and PLUMP, R. (1952): *AIEE Trans.* **71**, p. 348
CAZZANI, M., LISSANDRIN, M., MANGANARO, S., and MAZZA, G. (1978): 'Behaviour of EHV and UHV circuit breakers equipped with resistors during the making operation'. CIGRE Report 13-10 (Paris)
CHAPMAN, A. (1977): 'Electrical conductance of gas blast arcs'. Ph.D Thesis, University of Liverpool
CHAN, S.K., COWLEY, M.D., and FANG, M.T.C. (1976): 'Integral method for arc analysis. Pt 111', *J. Phys. D*, **9**, pp. 1085-1099
CHAN, S.K., FANG, M.T.C., and COWLEY, M.D. (1978): 'The DC arc in a supersonic nozzle flow', *IEEE Trans.*, **PS-6**, pp. 394-405
CHEN, D.C.C., and LAWTON, J. (1968): Rotary arc heaters *Trans. Inst. Chem Eng.*, **46**
CHINA, M. (1984): 'New methods of calculation for the production of circuit-breakers'. International Conference on Large high voltage electric systems, (CIGRE) Paper 13-09, pp. 1-4
CHRISTOPHOROU, L.G. (Ed.) (1980): 'Gaseous dielectrics' (Pergamon Press, Oxford)
CHU, F.Y., FORD, G.L., and LAW, C.K. (1982): 'Estimation of burn-through probability in SF$_6$ insulated substations', *IEEE Trans.*, **PAS-101**, pp. 1391-1399

CHU, F.Y., and TAHILIANI, V. (1980): 'Gas insulated substation fault survey', IEEE PES Winter Meeting, paper F80-226-1

CHU, F.Y., and LAW, C.K. (1980): 'Effects of power arc in gas insulated equipment'. Proc. 6th Int. IEE Conf. on Gas discharges and their applications, Edinburgh

CHU, F.Y., BOGGS, S.A, and LAW, C.K. (1980): Studies of power arc faults in SF_6 insulated equipment'. IEEE PES Winter Meeting, paper F80-225-3

CIGRE Working Group 13.02 (1980): 'Interruption of small inductive currents', *Electra*, (72), pp. 73–103

CIGRE Working Group 13.02 (1981): 'Interruption of small inductive currents', *Electra*, (72), pp. 5–30

COUSLEY, M.D. (1974): 'Integral method of arc analysis. Pt 1', *J. Phys. D*, **7**, pp. 2218–2231

CRAMERI, D., STOLARZ, W., BACHOFER, W., and ANGERMEIER, M. (1982): 'Extension of the 420 kV switching station at Lanpenburg by the addition of an SF_6 gas-insulated busbar', *Brown Boveri Rev.*, **6**, pp. 196–200

CUK, N., NISHIKAWARA, K.K., McCRAE, G.G., and ADAMS, P.T.B. (1980): 'Specification and application of SF_6 compressed gas insulated switchgear – a utility's point of view', *IEEE Trans.*, **PAS-99**, pp. 2241–50

DHAR, P.K., BARRAULT, M.R., and JONES, G.R. (1979): 'A multiring radio frequency technique for measuring arc boundary variations at high currents and close to current zero'. University of Liverpool, Arc Research Report, ULAP-T64

DIESSUER, A, and SCHRAMM, H.H. (1982): Discussion in Chu *et al.* (1982)

DUNCKEL, C., and VOIGT, R.P. (1985): 'Latest developments in the technology of the arc interruption and its application in MV and HV circuit breakers. New concept for HV switchgear installations'. Brown Boveri Publication CH-A 840315E

DUPLAY, C., and HENNEBERT, J. (1983): 'The breaking process with Fluarc SF_6 puffer and rotating arc circuit breakers and Rollarc contactors'. Merlin–Gerin Service Information E/CT 122

ECKLIN, G., ROBINSON, P., and SCHLICHT, D. (1982): 'Lightning overvoltage protection of the Drakensberg 420 kV SF_6 gas-insulated substation', *Brown Boveri Rev.*, **6**, pp. 188–195

EGGERT, H., GRIEGER, G., KOPPLIN, H., and LIPKEN, H. (1986): 'Operating technology for SF_6 puffer-breakers. Development, production and service experience'. Proc. Int. Conf. on Large high voltage electric systems, CIGRE paper 13-05

EIDINGER, A., and PETITPIERRE, R. (1979): 'The Brown Boveri range of switchgear', *Brown Boveri Rev.*, **4**, pp. 288–297

EL-AKKARI, F.R., and TUMA, D.T. (1977): 'Simulation of transient and zero current behaviour of arcs stabilized by forced convection', *IEEE Trans.*, **PAS-96**, pp. 1784–1788

ELIASSON, B., and SCHADE, E. (1977): 'Electrical breakdown of SF_6 at high temperature ($<2300\,k$)'. 13th Int. Conf. on Phenomena in ionised gases, Berlin, pp. 409–410

ELMCKE, B. (1981): 'High voltage technology', *Siemens Power Engng.*, **3** (Special Issue), pp. 4–10

ERICSSON, A. (1982): Discussion in Chu *et al.* (1982)

ERICSSON, A. (1982): Discussion in Chu *et al.* (1982)

FANG, M.T.C., RAMAKRISHNAN, S., and MESSERLE, H.K. (1980): 'Scaling laws for gas blast circuit breaker arcs during the high current phase', *IEEE Trans.*, **PS-8**, pp. 357–62

FANG, M.T.C., and BRANNEN, D. (1979): *IEEE Trans.*, **PS-7**, pp. 217–29

FANG, M.T.C. (1983): 'A review of gas blast circuit breaker arc modelling'. Liverpool University, Arc Research Report, ULAP-D10

FANG, M.T.C., BRANNEN, D., and JONES, G.R. (1978): 'The design of gas blast circuit breaker nozzles from a knowledge of arc properties'. Proc. IEE Int. Conf. on Developments in distribution switchgear, London, pp. 16–19

FANG, M.T.C., and NEWLAND, D.B. (1984): 'DC nozzle arcs with wall ablation'. University of Liverpool, Arc Research Report, ULAP-T69

FANG, M.T.C., RAMAKRISHNAN, S., and MESSERLE, H.K. (1980): 'Scaling laws for gas blast circuit breaker arcs during the high current phase', *IEEE Trans.*, **PS-8**, pp. 357–62

References 183

FAWDREY, C.A. (1979): 'SF$_6$ circuit breaker mechanisms'. Proc. Int. Conf. on EHV systems, London, pp. 36–41
FLURSCHEIM, C.H. (1982): 'Power circuit breaker theory and design' (Peter Peregrinus, revised edn.)
FRIND, G.: 'Experimental investigation of limiting curves for current interruption of gas blast breakers'. in RAGALLER, K. (Ed.) (1978); 'Current interruption in high voltage networks' (Plenum Press, NY) pp. 67–94
FRIND, G., KINSINGER, R.E., MILLER, R.D., NAGAMATSU, H.T., and NOESKE, H.O. (1977): 'Fundamental investigations of arc interruption in gas flows'. EPRI Final Report EL-284
FRIND, G., and RICH, J.A. (1974): 'Recovery speed of axial flow gas interrupter: Dependence on pressure and di/dt for air and SF^6', *IEEE Trans.*, **PAS-93**, pp. 1675–1684
FROST, L., and LIEBERMAN, R.W. (1971): 'Composition and transport, properties of SF$_6$', *Proc. IEEE*, Vol. 59, p. 4
GABIN, J., GRATZMULLER, J., GAILLY, M., LACOSTE, A., and LORCET, P. (1974): 'Description and tests of a high voltage current circuit breaker', *Rev. Gen. Elect.*, **83**, pp. 561–9
GARRAD, C.J.O. (1976): 'High voltage switchgear', *Proc. IEE.*, **123**, pp. 1053–1080
GARZON, R.D. (1976): *IEEE Trans.*, **PAS-95**, p. 1681
GAYET, P., BERNARD, G., DEMOUTUSSAINT, D., and LALOT, J. (1984): 'Research into a voltgae test procedure for site testing metalclad substations with maximum efficiency', CIGRE (Paris), Paper 23-05
'SF$_6$ switchgear for transmission and distribution'. GEC High Voltage Switchgear Publication 1294-74
GRANT, D.M., PERKINS, J.F., CAMPBELL, L.C., IBRAHIM, O.E., and FARISH, O. (1976): 'Comparative interruption studies of gas blasted arcs in SF$_6$-H$_e$ mixtures'. Proc. 4th Int. Conf. Gas Discharges, Swansea. IEE Conf. Publ. 143
GREENWOOD, A.N., BARKAR, P., and KRACHT, W.C., (1972): 'HVDC vacuum circuit breakers', *IEEE Trans.*, **PAS-91**, (4), pp 1575–88
HARRIS, A.F.W. (1984): 'Progress in development and research for EHV circuit breaker interrupters'. BEAMA Indonesia Seminar for PLM (Jakarta)
HARRIS, A.F., and SIMMS, J.R. (1978): 'The development of a 420 kV metalclad switch disconnector'. Int. Conf. on EHV Systems, London. IEE Conf. Publ. 168, pp. 14–19
HEADLEY, A. (1984): 'Meeting system reqrueiments with modern switchgear'. Proc. IEE Symp. on Trends in modern switchgear design 3·3–150 kV, Newcastle, pp. 9.1–9.5
HERMANN, W., KOGELSCHATZ, U., NIEMEYER, L., RAGALLER, K., and SCHADE, E. (1976): *IEEE Trans.*, **PAS-95**, pp. 1165–1176
HERMANN, W., and RAGALLER, K. (1979): 'Development tests for circuit breakers'. *Brown Boveri REv.*, **4**, pp. 281–287
HERMANN, W., and RAGALLER, K. (1977): 'Theoretical description of the current interruption in HV gas blast circuit breakers', *IEEE Trans.*, **PAS-96**, pp. 1546–1555
HEY, E.N., and PRICE, J.F. (1982): *J. Physiol* (London), **330**, p. 429
HOEGG, P. (1980): Dicussion in Cuk *et al.* (1980)
HOFMANN, G.A., LA BARBERA, G.L., READ, N.E., and SHILLONG, L.A. (1976): 'A high speed HVDC circuit breaker with cross-field interrupters', *IEEE Trans.*, **PAS-95**, p. 1182
HORIUCHI, T., YANABU, S., TAMAGAWA, T., NISHIWAKI, S., and TOMMIMURO, S. (1980): 'Development of full-scale high-voltage DC circuit breaker'. Conf. on Incorporating HVDC power transmission into system planning, Phoenix, Arizona
HURLEY, J.R. (1973): 'Magnetic forces on gas blast interrupter arcs due to electrode shape'. IEEE PES Winter Meeting, paper T73, 058-5, pp. 1525–1530
HUMPHRIES, M.B., LUGTON, W.L., and FAWDREY, C. (1977): 'Statistical approach to testing and application of circuit breakers'. CIGRE SC 13, Poznan
IKEDA, H., UEDA, T., KOBAYASHI, A., YAMAMOTO, M., and YANABU, S. (1984): 'Development of large-capacity, SF$_6$ interruption chamber and its application to GIS', *IEEE Trans.*, **PAS-103**, pp. 3038–3043
IKEDA, S., AOYAGI, A., and AMEMIYA, T. (1981): 'Diagnostic technique for mechanical

failures of gas circuit breakers'. IEEE PES Summer Meeting, Portland, paper 81 SM 471-2, pp. 1–7

ISHIKAWA, M., IKEDA, H., YANABU, S., and YAMAMOTO, M. (1984): 'Numerical study of delayed-zero-current interruption phenomena using transient analysis model for an arc in SF_6 flow'. IEEE PES Trans. and Distrib. Conf., Kansas City, paper 84 T&D 330-7, pp. 1–7

JAKOB, T., SCHADE, E., and SCHAUMANN, R. (1985). 'Self-extinction, and economic new principle for SF_6 circuit breakers'. Brown Boveri Publication CH-A 194 280E

JONES, G.R.: 'Current interruption in high voltage networks'. in RAGALLER, K. (Ed.) (1978): 'The influence of turbulence on current interruption' (Plenum Press)

JONES, H.A., DAVIES, E.E. and HUGHES, J.M. (1982): *Bull. Eur. Physiopathol. Respir.*, **18**, p. 391

JONES, G.R. (1988): 'Electric arcs in industrial devices – diagnostic and experimental techniques' (Cambridge University Press)

JONES, G.R. (1983): 'High current arcs at high pressures'. Proc. 16th Int. Conf. on Phenomena in ionized in gases, Dusseldorf, invited papers pp. 106–18

JONES, G.R. (1984): 'Predicting pressure transients due to arcing in two pressure, puffer and rotary arc interrupters'. Proc. IEE Symp on Trends in modern switchgear design 3·3–150 kV, Newcastle, pp. 8.1–8.5

JONES, G.R., and FANG, M.T.C. (1980): 'The physics of high power arcs', *Rep. Prog. Phys.*, **43**, pp. 1415–1465

JONES, G.R., LECLERC, J.L., and SMITH, M.R. (1982): 'Self magnetic effects in a model gas blast circuit breaker at very high currents', *IEE Proc.*, **129A**, pp. 611–618

JONES, G.R., TURNER, D.R., CHEN, D., PARRY, J., and SPENCER, J. (1986): 'Factors affecting the performance and properties of an SF_6 rotary arc interrupter'. Proc. 2nd Int. Conf. on Developments in distribution on switchgear

JONES, G.R., SHAMMAS, N.Y., and PRASAD, A.N. (1986): 'Radiatively induced nozzle ablation in high-power circuit interrupters', *IEEE trans.*, **PS-14**, pp. 413–422

KING, L.A.H. (1978): 'Research studies of arc interruption in SF_6 circuit breakers', *J. Sci. & Technol.*, **45**, p. 23

KINSINGER, R.E., and NOESKE, H.O. (1976): 'Relative arc thermal recovery speed in different gases'. Proc. IEE 4th Int. Conf. on Gas Discharges, (Swansea), pp. 24–28

KIRCHESCH, P., and NIEMEYER, L. (1985): 'Arc behaviour in an ablating nozzle'. 5th Int. Symp. on Switching arc phenomena, Lodz, Poland, pp. 39–43

KOBAYASHI, A., YANABU, S., YAMASHITA, S., TOMIMURO, S., and HAGINOMORI, E. (1978): 'Measurement of voltage distribution phenomena in SF_6 single pressure circuit breaker', *IEEE Trans.*, **PAS-97**, pp. 1–9

KOEPPL, G., STEPINSKI, B., FREY, H., and KOLBE, W. (1983): 'The crossed-ring arrangement: A new concept for HV switchgear installations', *IEEE Trans.*, **PAS-102**, pp. 355–363

KOPAINSKY, J.: 'Breakdown in circuit breakers'. in RAGALLER, K. (Ed.) (1979): 'Current interruption in high voltage networks' (Plenum Press, New York) pp. 329–354

KOPAINSKY, J., and MURI, K. (1983): 'A fault arc detection system for closed spaces', *Brown Boveri Rev.*, Jan/Feb, pp. 79–80

KOPAINSKY, J., and SCHADE, E. (1979): *Appl. Phys.*, **20**, pp. 147–153

KOVITYA, P., LOWKE, J.J., and STOKES, A.D. (1978): Proc. Elect. Engng. Conf. Canberra EE14, pp. 220–224

KULICKE, B., and SCHRAMM, H.H. (1980): 'Clearance of short-circuit with delayed current zeros in the Itaipu 550 kV substation, *IEEE Trans.*, **PAS-99**, pp. 1406–1414

KURIMOTO, A. (1985): 'Factors affecting the arc interruption of SF_6 puffer type breakers'. Proc. Int. Conf. on Gas discharges and their applications, Oxford (Leeds University Press)

KUWAHARA, H., YOSHINAGA, K., SAKUMA, S., YAMAUCHI, T., and MIYAMOTO, T. (1982): 'Fundamental investigation on internal arcs in SF_6 gas-filled enclosures', *IEEE Trans.*, **PAS-101**, pp. 3977–3987

KUWAHARA, H., TANABE, T., SASAMOTO, S., OGAWA, A., and NITTA, T. (1983*a*):

References 185

'Experiences in the internal inspections and maintenance works on gas insulated equipment in the field', *IEEE Trans.*, **PAS-102**, pp. 843–851

KUWAHARA, H., TANABE, T., YOSHINAGA, K., SAKUMA, S., IBUKI, K., and YAMADA, K. (1983b): 'Investigation of dielectric recovery characteristics of hot SF_6 gas after current interruption for developing new 300 kV and 550 kV gas circuit breakers'. IEEE PES Summer Meeting, Los Angeles, paper 83 SM 502-2, pp. 1–6

KUWAHARA *et al.* (1980): 2nd Int. Symp. on Gaseous dielectrics, Knoxville, TN (Pergamon Press)

LECLERC, J.L., SMITH, M.R., and JONES, G.R. (1980) 'Pressure transients in a model gas blast circuit breaker operating at extra high current levels'. *IEEE Trans* **PS-8**, pp. 376–384

LEE, A. (1982): 'Arc-circuit instability. HVDC circuit breaker concept based on SF_6 puffer technology'. Proc. 7th Int. Conf. on Gas discharges and their applications, London, paper A5

LEE, A., and FROST, L.S. (1980): 'Interruption capability of gases and gas mixtures in a puffer type interrupter', *IEEE trans.*, **PS-8**, p. 362

LEE, A., SLADE, P.G., YOON, K.H., PORTER, J., and VITHAYATHIL, J. (1985): 'The Development of an HVDC SF_6 breaker'. IEEE PES Winter Meeting, New York, paper 85 WM 248-0

LEUPP, P. (1983): 'SF_6 circuit breakers for outdoor and GIS installations at voltages up to 800 kV', *Brown Boveri Rev.*, **1**, pp. 53–57

LEWIS, E., SHAMMAS, N.Y., and JONES, G.R. (1985): 'The current zero SF_6 blast arc at high di/dt'. Proc. Int. Conf. on Gas discharges and their applications, Oxford (Leeds University Press)

LIEBERMANN, R.W., and LOWKE, J.J. (1976): *J. Quant. Spectrosc. Radiat. Transf.*, **16**, pp. 263–264

LIPS, H.P. (1980): 'Prospects for multiterminal HVDC transmission'. IEEE PES Winter Meeting

LISTER, C.A. (1984): 'Vacuum, SF_6 and air-break contactors for medium voltage controllers', *IEEE Trans.*, **PAS-103**, pp. 3021–3029

LOWKE, J.J., and LUDWIG, H.C. (1975): 'A simple model for high current arcs stabilised by forced convection', *J. Appl. Phys.*, **46**, p. 3352

LOWKE, J.J. (1974): 'Nett radiative emission calculations' *J. Quant. Spect. Rad. Transf.*, **14**, pp. 111–22

LOHMANN, V.W., and BOLTON, A.C. (1985): 'Gas insulated switchgear developed for 765 kV grid', *Modern Power Systems*, pp. 29–33

LUTZ, F., and PIETSCH, G. (1982): 'The calculation of overpressure in metal enclosed switchgear due to internal arcing', *IEEE Trans.*, **PAS-101**, pp. 4230–4236

MAECKER, H.H. (1955): 'Plasmalromungen in Lichtbogen infolge eigen magnetischer Kompression'. *Z. Phys.*, **141**, pp. 198–216

MANGANARO, S., and ROVELLI, S. (1977): 'A new circuit for synthetic auto-reclosing test duties under short circuit conditions on high power circuit breakers'. IEEE PES Winter Meeting, New York, Paper F77 132-4

MANGANARO, S., and SCHRAMM, H.H. (1980): 'Application of synthetic auto-reclosing circuit for testing high-voltage circuit-breakers', *IEEE Trans.*, **PAS-99**, pp. 2223–2231

MAYER, A. (1983): 'Surge arresters for limiting overvoltages in high-voltage systems', *Brown Boveri Rev.*, **1**, pp. 48–52

MAYER, J.C., and NAGAMATSU, H.T. (1980): 'Calculation of upsteam flow field for single flow interrupter nozzles', GE Research and Development, Class 1 TIS Report, no. 80CRD032, Schenectady

MAZZA, G., and MICHACA, R. (1980): 'The first international enquiry on circuit breaker failures and defects in service', *Electra*, (79), pp. 22–89

MITSUBISHI: SF_6-gas generator breaker type SFWA, HBN-84785

MIYACHI, I., and NAGANAWA, H. (1947): 'Spiral arc in SF_6 facilitating DC interruption'. Proc. 3rd Int. Conf. on gas discharges, (London), IEE Conf. Publ. 118, pp. 521–524

MOLL, R., and SCHADE, E. (1979): 'Investigation of the dielectric recovery of SF_6 blown high

186 References

voltage switchgear arcs'. Proc. 3rd Int. Symp. High voltage engineering, Milano, Italy, paper 32.07

MOLL, R., and SCHADE, E. (1980): 'Dielectric recovery of axially blown SF_6 arcs'. Proc. 6th Int. Conf. on Gas discharges and their applications, Edinburgh, IEE Conf. Publ. 189, Pt. I, p. 13

MORII, K., MATSUMURA, S., OKADA, T., USHIO, T., TOMINAGA, S., and INAMURA, S. (1978): 'Field experience and future trends of 500 kV gas insulated metalclad substations'. Int. Conf. on Large high voltage electric systems (CIGRE) Pt. 1, pp. 23-02/1–20

MORII, K., MATSUMURA, S., USHIO, T., TOMINAGA, S., and SHINOZAKI, Y. (1979): 'SF_6 gas circuit breaker; new solution for generator main circuit switching', *IEEE Trans.*, **PAS-98,** pp. 759–769

MORITA, T., TSUTSUMI, M., SHIGA, S., IWAI, H., and MIYACHI, I. (1985): 'New breaking test methods for circuit-breakers and disconnectors of 3-phase enclosed GIS'. Proc. Int. Conf. on Gas discharges and their applications, Oxford

MURAI, Y., YAMAJI, S., MIYAMOTO, T., SASAO, H., and UEDA, Y. (1981): An improvement of low current interrupting capability in self-interruption GCB'. IEEE PES Summer Meeting, Portland, paper 81 SM 413-4, pp. 1–5

MUNRANO, M., FUJII, T., and YAMASHITA, S. (1974): 'Synthetic test of high power circuit breaker'. IEEE Conf. Record of Symp. on High power testing, pp. 85–90

MURASE, H., OHSHIMA, I., AOYAGI, H., and MIWA, I. (1985): 'Measurement of transient voltage induced by disconnect switch operation', *IEEE Trans.*, **PAS-104,** pp. 157–165

NAGANAWA, H., OHNO, H., IIO, M., and MIYACHI, I. (1985): 'DC interruption by spiral arc in SF_6–N_2 mixture'. Proc. Int. Conf. on Gas discharges and their applications, Oxford (Leeds Universuty Press)

NAGATA, M., MIYACHI, I., YOKOI, Y., and ISAKA, K. (1980): 'Breakdown characteristics of high temperature air and SF_6 gas'. Proc. 6th Int. Conf. on Gas discharges, IEE Conf. Publ. 189, pp. 78–81

NAKAGAWA, Y., TSUKUSHI, M., HIRASAWA, K., and YOSHIOKA, Y. (1985): 'Transient dielectric recovery characteristics across the contacts after current interruption of puffer type gas circuit breakers'. Proc. Int. Conf. on Gas discharges and their applications, Oxford

NAKANISHI, K., ISHIKAWA, M., MATSUMURA, S., TERANISHI, Y., TOMINAGA, S., YANABU, S., and HIRASAWA, K. (1982): 'Verification test of tank-type gas circuit-breakers considering the actual stresses expected in the field'. Int. Conf. on Large high voltage electric systems, pp. 1–9

NATSUI, K., YOSHIOKA, Y., and HIRASAWA, K. (1977): 'Nozzle clogging phenomena in puffer type gas circuit breaker'. 3rd Int. Symp. on Switching arc phenomena, Poland, pp. 172–176

NATSUI, K., NAKANURA, I., KOYANAGI, O., YOSHIOKA, Y., and HIRASAWA, K. (1980): 'Experimental approach to one-cycle puffer type SF_6 gas circuit breaker', *IEEE Trans.*, **PAS-99,** pp. 833–40

NIEMEYER, L. (1978): 'Evaporation-dominated high current arcs in narrow channels', *IEEE Trans.*, **PAS-97,** pp. 950–958

NIEMEYER, L., and RAGALLER, K. (1973): 'Development of turbulence by the interaction of gas flow with plasmas', *Z. Naturf.*, **28a,** p. 1281–1289

NISHIWAKI, S., KANNO., Y., SATO, S., HAGINOMORI, E., YAMASHITA, S., and YANABU, S. (1983): 'Ground fault by restriking of SF_6 gas-insulated disconnecting switch and its synthetic tests', *IEEE Trans.*, **PAS-102,** pp. 219–2216

NITTA, T., SHIBAYA, Y., and FUJIWARA, Y. (1975): 'Voltage-time characteristic of electrical breakdown in SF_6', *IEEE Trans.*, **PAS-94,** pp. 105–115

NOBLE, T.J. (1984): 'Gas insulated vacuum switchgear up to 36 kV'. Proc. IEE Symp. on Trends in modern switchgear design 3·3–150 kV, Newcastle, pp. 3.1–3.5

NOESKE, H.O. (1977): 'Investigation of dynamic nozzle parameters for various nozzle geometries and test conditions of an experimental half-size SF_6 puffer breaker', *IEEE Trans.*, **PAS-96,** pp. 896–906

References

NOESKE. H.O. (1981): 'Arc thermal recovery speed in different gases and gas mixtures', *IEEE Trans.*, **PAS-100,** pp. 4612–462–

NOESKE, H.O., BENENSON, D.M., FRIND, G., HIRASAWA, K., KINSINGER, R.E., NAGAMATSU, H.T., SHEER, R.E., and YOSHIOKA, Y. (1983): 'Applications of arc-interruption fundamentals to nozzles for puffer interrupters', EPRI Report EL-3293, Project 246-2

OAKES, M.C. (1986): 'SF_6 modular distribution switchgear – a logical solution'. 2nd Int. IEE Conf. on Developments in distribution switchgear, London, pp. 53–58

OAKES, M.C. (1984): 'A logical approach to the design of distribution switchgear for tropical applications'. Inex–Asia Conference, Singapore

OGAWA, S., HAGINOMORI, E., NISHIWAKI, S., YOSHIDA, T., and TERASAKA, K. (1985): 'Estimation of restriking transient overvoltage on disconnecting switch for GIS'. IEEE Summer Meeting, paper 85 SM 367-8

OHSHIMA, I., KOJIMA, S., and UESAKA, Y. (1983): 'Insulation co-ordination of 500 kV GIS'. Canadian Electrical Association, Spring Meeting, Vancouver, Canada, pp. 1–12

OLSEN, W., and RIMPP, F. (1981): 'SF_6 insulated high-voltage switchgear: Present status and development trends', *Siemens Power Enging.*, **3,** (Special Issue on High Voltage Technology), pp. 36–41

PARRY, J. (1983): 'A new rotating arc SF_6 circuit breaking element with a wide range of application in the field of distribution switchgear'. CIRED (Liege), pp. e05.1–e05.5

PARRY, J. (1984): 'Further developments in SF_6 switchgear for distribution systems incorporating a rotating arc circuit breaking device'. Proc. IEE Symp. on Trends in modern switchgear design 3·3–150 kV, Newcastle, pp. 2.1–2.6

PERRY, M. (1981): Discussion on Tominaga *et al.* (1981)

PETTERSSON, K.G., and GRAUSTROM, E.V. (1977): 'Design of gas insulated substations with respect to internal arcs'. IEE Conf. Publ. 157, pp. 66–70

PRYOR, B.M. (1984): 'HV fuse-switch combinations'. IEE NE Centre, Symp. on Trends in modern switchgear design 3·3–150 kV, Newcastle, pp. 16.1–16.4

RAGALLER, K. (1974): *Z. Naturf.*, **29a,** pp. 556–567

RAGALLER, K., EGLI, W., and BRAND, K.P. (1982): 'Dielectric recovery of an axially blown SF_6-arc after current zero: Part II Theoretical investigations', *IEEE Trans.*, **PS-10,** pp. 154–162

RIEDER, W., and URBANEK, J. (1966): 'New aspects of current zero research on circuit breaker reignition. A theory of non equilibrium arc conditions'. CIGRE Paper 107

RONDEEL, W, and WIESSFERDT, P. (1984): 'Transformer protection using SF_6-insulated ring main units', Proc. IEE Symp. on Trends in modern switchgear design 3·3–150 kV, Newcastle, pp. 14.1–14.7

RUCHTI, C.B. (1985): 5th Int. Symp. on Switching arc phenomena, Lodz, Poland, pp. 34–38

RUFFIEUX, S., HENNEBERT, J., and ROMIER, R. (1986): 'Evolution of operating mechanisms for 7·2 to 36 kV SF_6 circuit breakers related to progress in breaking techniques'. Proc. Int. Conf. on Large high voltage electric systems. (CIGRE), paper 13-03

RUOSS, E. (1979): 'Overvoltages on energizing HV lines', *Brown Boveri Rev.*, **4,** pp. 262–270

RYAN, H.M., ALI, S.M.G., and POWELL, C.W. (1983): 'Field computation relating to switchgear design'. Proc. ISH Conference (Athens)

RYAN, H.M., MILNE, D., and POWELL, C.W. (1985): 'Site testing and the evaluation of a technique to differentiate between a gas or spacer flashover in SF_6 GIS'. Symposium on gas-insulated substations technology and practice, pp. 1–8

RYAN, H.M., and WATSON, W.L. (1978): 'Impulse breakdown characteristics in SF_6 for non uniform field gaps'. CIGRE Paper 15.01

SACKNER, M.A., RAO, A.S., BIRCH, S., ATKINS, N., GIBBS, L., and DAVIS, B. (1982): *Chest,* **83,** p. 137

SAKAI, M., KATO, Y., TOKUYAMA, S., SUGAWARA, H., and ARIMATSU, K. (1981): 'Development and field application of metallic return protecting breaker for HVDC transmission', *IEEE Trans.*, **PAS-100,** (12)

SAITO, K., and HONDA, H. (1982): 'Recent SF_6 gas circuit breaker developments', *Hitachi Rev.*, **31,** pp. 151–156

188 References

SASAO, H., HAMANO, S., OOMORI, T., UEDA, Y., and MURAI, Y. (1981): 'Mixing process of arced gas with cold gas in the cylinder of gas circuit breaker'. IEEE PES Summer Meeting, paper 81 SM, pp. 3440–3449

SASAO, H., HAMANO, S., UEDA, Y., YAMAJI, S., and MURAI, Y. (1982): 'Dynamic behaviour of gas-blast arcs in SF_6-N_2 mixture', *IEEE trans.*, **PAS-101**, pp. 4024–4029

SATYANARAYANA, P., and BRAUN, D. (1984): 'Applications of the SF_6 self-extinguishing circuit breaker for switching small inductive currents'. Indian Electrical Manufacturers' Associated Int. Seminar on Switchgear and Control Gear, Bombay, Session IV, pp. 35–48

SAUERS, I., ELLIS, H.W., CHRISTOPHOROU, L.G., GRIFFIN, G.D., and EASTERLY, C.E. (1984): 'Spark decomposition of SF_6; toxicity of byproducts'. Oak Ridge National Laboratory Report ORN/TM-9062

SCHADE, E. (1985): 'Similarity of the dielectric recovery characteristic of axially blown arcs in SF_6'. Proc. Int. Conf. on Gas discharges and their applications, Oxford (Leeds University Press)

SCHADE, E. (1985): 'Similarity of the dielectric recovery characteristic of axially blown arcs in SF_6'. Proc. Int. Conf. on Gas discharges and their applications, Oxford (Leeds University Press)

SCHADE, E., and RAGALLER, K. (1982): 'Dielectric recovery of an axially blown SF_6 arc after current zero. Part 1 Experimental Investigations', *IEEE Trans.*, **PS-10**, pp. 141–153

SCHMITZ, W., and KLINK, P.J. (1978): 'A special design of SF_6 metalclad switchgear for HV distribution systems'. Conf. on Distribution switchgear, IEE Publ. 168, pp. 36–40

SCHRADE, H.O. (1973): 'Stable configuration of electric arcs in transverse magnetic fields', *IEEE Trans.*, **PS-1**, pp. 47–54

SCOTT, H.F., MATTINGLEY, J., and RYAN, H.M. (1974): 'Computation of electric fields: Recent developments and practical applications', *IEEE Trans.*, **El-9**, pp. 18–25

SENDA, T., TAMAGAWA, T., HIGUCHI, K., HORIUCHI, T., and YANABU, S. (1983): 'Development of HVDC circuit breaker based on hybrid interruption scheme'. IEEE PES Sumer Meetimg, Paper 83 SM 501-4, pp. 1–7

SHIMMIN, D.W. (1986): 'High power short-circuit studies on an SF_6 puffer circuit breaker'. Ph.D. Thesis, Univ. of Liverpool

SHIMMIN, D.W., ALI, S.M.G., HEADLEY, P., RYAN, H.M., and JONES, G.R. (1985): 'A comparison of the thermal performance of two pressure and puffer circuit breakers'. Proc. Int. Conf. on Gas discharges and their applications, Oxford (Leeds University Press)

SHISHKIN, G.G., and JONES, G.R. (1985): 'Electromagnetic emission from high current, convection controlled arcs'. Proc. 8th Int. Conf. on Gas discharges and their applications, Oxford (Leeds University Press)

SIMMS, J.R. (1978): IEE Int. Conf. on HV switchgear, pp. 14–19

SIMMS, J.R. (1984): 'Overvoltage studies in gas insulated switchgear'. IEE NE Centre, Symp. on Trends in modern switchgear design 3·3–150 kV, Newcastle, pp. 6.1–6.6

SMITH, M.R., LECLERC, J.L., and JONES, G.R. (1981): 'The electrical characteristics of gas blast circuit breaker arcs at very high currents'. IEEE PES Summer Meeting, Portland, Oregon, paper 81 SM 409-2

SMITH, M.R., and JONES, G.R. (1981): 'The protection of a current zero shunt from overload at high peak currents'. University of Liverpool, Arc Research Report ULAP-T68

SPENCER, M.C., and HARRIS, M.C. (1978): 'A new digital recorder timer for multi-gap high voltage circuit breakers'. IEE Int. Conf. on Development of distribution switchgear, London, pp. 130–134

SPENCER, J., PARRY. J., and JONES, G.R. (1985): 'Complex aspects of arc behaviour in a rotary arc circuit breaker'. Proc. Int. Conf. Gas discharges and their applications, Oxford (Leeds University Press)

SPINDLE, H.E. (1980): Discussion in Cuk *et al.* (1980)

STRACHAN, D.C., LIDGATE, D., and JONES, G.R. (1977): 'Radiative energy losses from a high current air blast arc", *J. Appl. Phys.*, **48**, pp. 2324–2330

STRASSER, H., SCHMIDT, K.D., and HOGG, P. (1975): 'Effects of arcs in enclosures filled with SF_6', *IEEE Trans.*, **PAS-94**, pp. 1051–1060

References

STEPHANIDES, H.V., AESBACK, B., and SCHOETZAU, H.J. (1986): 'Modern methods for the reduction of the operating energy for SF_6 circuit-breakers'. Proc. Int. Conf. on Large high voltage electric systems, (CIGRE), paper 13-12

STEPINSKI, B., (1978): 'New aspects of the construction of outdoor switchgear installations for voltages up to 765 kV', *Brown Boveri Rev.*, **65**, pp. 268–275

STEWART, J.S. (1979): 'SF_6 circuit-breaker design and performance', *Electron. & Power*, **25**, pp. 121–126

STEWART, J.S. (1984): 'An autorecloser with microprocessor control for overhead line distribution', *Electron. & Power*, **30**, pp. 469–472

STOKES, A.D. (1976): Proc. IEE Int. Conf. on Gas discharges, Swansea, pp. 75–78

SUZUKI, K., MIZOGUCHI, H., SHIMOKAWARA, N., MURAYANNA, Y., and YANABU, S. (1984): 'Current interruption by disconnecting switch and earthing switch in GIS', *IEE Proc. c*, **131**, pp. 54–60

SUZUKI, T., YOSHIDA, H., KOYAMA, A., and TOMIMURO, S. (1982): 'Degradation process of grease due to SF_6 gas discharge products', *IEEE Trans.*, **PAS-101**, pp. 2805–2809

SWANSON, B.W. (1977): 'Nozzle arc interruption in supersonic flow', *IEEE Trans.*, **PAS-96**, pp. 1697–1706

SWANSON, B.W., and ROIDT, R.M. (1971): 'Some numerical solutions of the boundary layer equations for an SF_6 arc', *Proc. IEEE*, **59**, pp. 493–501

SWANSON, B.W., and ROIDT, R.M. (1972): 'Thermal analysis of an SF^6 circuit breaker arc', *IEEE Trans.*, **PAS-91**, pp. 381–389

SWANSON, B.W., ROIDT, R.M., and BROWNE, T.E. (1972): 'A thermal arc model for short line fault interruption', *ETZ A*, **93**, pp. 375–380

SZENTE-VARGA, H.P., and TECCHIO, P. (1983): 'SF_6 gas-insulated switchgear for all applications', *Brown Boveri Rev.*, **1**, pp. 81–86

TAHIR, (1978): 'Arc modelling using the integral method'. M.Eng. thesis, University of Liverpool

TAKAHASHI, I., YOSHIOKA, Y., HIRASAWA, K., and TSUKISHI, M. (1974): 'Research on a new type magnetic puffer gas circuit breaker, *IEEE Trans.*, **PAS-97**, pp. 421–428

TANDON, S.C., ARMOR, A.F., and CHARI, M.V.K. (1983): 'Nonlinear transient finite element field computation for electrical machines and devices', *IEEE Trans.*, **PAS-102**, pp. 1089–1095

TAYLOR, S. (1983): 'Data acquisition and analysis of circuit breaker are measurements'. M.Eng. thesis, University of Liverpool

THURIES, E., JEANJEAN, R., and VAN DOAN, P. (1980): 'Technico-economical response of the generator circuit breaker to future generation ratings', *IEEE Trans.*, **PAS-99**, pp. 1970–1974

TODOROVIC, P.S., and JONES, G.R. (1985): 'Time resolved laser Doppler measurements in the underexpanded jet of a model gas blast circuit breaker under arcing conditions', *IEEE Trans.*, **PS-13**, pp. 153–62

TOKORO, K., HARUMOTO, Y., YAMAMOTO, H., YOSHIDA, Y., MUKAE, H., OHNO, Y., SHIMADA, M., and IDA, Y. (1982): 'Development of electronic potential and current transducers suitable for gas insulated switchgear and adequate for application to substation digital control system', *IEEE Trans.*, **PAS-101**, pp. 3967–3975

TOKUYAMA, S., ARIMATSU, K., YOSHIOKA, Y., KATO, Y., and MIRATA, R. (1985): 'Development and interrupting tests on 250 kV 8 kA HVDC circuit breakers'. IEEE PES Winter Meeting, New York, paper 85 WM 250-6

TOMINAGA, S., *et al.* (1979): 'Estimation and performance investigation on SLF interrupting ability of puffer type gas circuit breaker', *IEEE Trans.*, **PAS-98**, pp. 261–269

TOMINAGA, S., KUWAHARA, H., HIROOKA, K., and YOSHIOKA, T. (1981): 'SF_6 gas analysis technique and its application for evaluation of internal conditions in SF_6 gas equipment', *IEEE Trans.*, **PAS-100**, pp. 4196–4206

TRUHAUT, R., BOUDENE, C., and DUET, J.L. (1973): *Arch. Mal. Prof.*, **34**, p. 581

TUMA, D.T. (1980): 'A comparison of the behaviour of SF_6 and N_2 blast arcs around current zero', *IEEE Trans.*, **PAS-99**, pp. 2129–2137

190 References

TUMA, D., and LOWKE, J.J. (1975): 'Prediction of properties of arcs stabilised by forced convection', *J. Appl. Phys.,* **46,** pp. 3361–7

TURNER, D.R., and CHEN. D. (1983): 'The transient magnetic field analysis of a rotating arc switch'. Proc. Univ. Power Electrial Conf. (Huddersfield) pp. xxx

UEDA, Y., MURAI, Y., OHNO, A., and TSUTSUMI, T. (1982): 'Development pf 7.2 kV – 63 kA advanced puffer gas circuit breaker', *IEEE Trans.,* **PAS-101,** pp. 1504–1510

UEDA, Y., SASAO, H., MURAI, Y., MIYAMOTO, T., and ITOH, T. (1979): 'Arcing phemomena in puffer breakers'. Symp. on High voltage switchgear equipment, Sydney, Australia

UEDA, Y., SASAO, H., MURAI, Y., YOSHIMAGA, K., MIYAMOTO, T., and TOMINAGA, S. (1981): 'Self-flow generation phenomena in a gas circuit breaker without puffer action'. IEEE PES Winter Meeting, Paper No. 81 WM 149-4

UHLENBUSCH, J. (1976): *Physica,* **82c,** pp. 61–85

USHIO, T., SHIMURA, I., KUWAHARA, H., and YOSHINAGA, K. (1972): 'Consideration of transient recovery voltages in connection with the modern circuit breaker testing', *IEEE Trans.,* **PAS-91,** pp. 733–40

USHIO, T., TOMINAGA, S., KUWAHARA, H., MIYAMOTO, T., UEDA, Y., and SASAO, H. (1981): 'SLF interruption by a gas circuit breaker without puffer action'. IEEE PES Winter Meeting, Atlanta, GA, paper 81 WM 510-2

VAN DER LINDEN, W.A., and VAN DER SLUIS, L. (1983): 'A new artificial line for testing high-voltage circuit breakers', *IEEE Trans.,* **PAS-102,** pp. 797–803

VITHAYATHIL, J.J. (1983): 'HVDC breaker and its application'. Int. Conf. on HVDC technology, Rio de Janeiro

VITHAYATHIL, J.J., COURTS, A.L., PETERSON, W.G., HINGORANI, N.G., NILSSON, S., and PORTER, J.W. (1985): 'HVDC circuit breaker development and field tests'. IEEE PES Winter Meeting, New York, paper 85 WM 252-2

WALKER, R. (1984): 'Responding to trends – a user viewpoint'. IEE NE Centre Conf. on Trends in modern switchgear design 3·3–150 kV, pp. 17.1–17.2

WALMSLEY, H.L., and JONES, G.R. (1980): 'Correlation of the radially integrated properties of gas blast are discharges', *IEEE Trans.,* **PS-8,** pp. 39–49

WANG, B., TAYLOR, S., BLACKBURN, T.R., and JONES, G.R. (1982): 'Thermal reignition performance limitations of a model SF_6 circuit breaker under full and scaled power conditions'. University of Liverpool, Arc Research Report ULAP-T72

WIELAND, A., KOGLIN, B.B., and LEONHARDT, G. (1982): 'Package substation system type Enk with SF_6 insulation', *IEEE Trans.,* **PAS-101,** pp. 2300–2307

WITTLE, J.K., and HOUSTON, J.M. (1982): 'Fault detection sensors for gas insulated equipment'. EPRI Report EL-1149

YAMAMOTO, M., and HONDA, M. (1982): 'Testing facilities for developing UHV equipment', *IEE Trans.,* **PAS-101,** pp. 2314–2318

YAMAMOTO, M., YAMASHITA, S., IKEDA, H., and YANABU, S. (1985): 'New short-circuit testing facilities to cope with the recent development of GIS', *IEEE Trans.,* **PAS-104,** pp. 150–156

YAMASHITA, S., MIYAKE, N., TOMIMURO, S., YANABU, S., and KOBAYASHI, A. (1978): 'A method of generating four-parameter transient recovery voltage during synthetic tests of large capacity circuit breakers', *IEEE Trans.,* **PAS-97,** 1–9

YANABU, S., OHISHI, M., OZAKI, Y., and MURAYAMA, Y. (1985): 'Development of UHV gas circuit breaker and disconnecting switch and their performance switch'. Private communication

YANABU, S., MIZOGUCHi, H., KOBAYASHI, A., OZAKI, Y., and MURAKAMI, Y. (1981): 'Factors influencing the interrupting ability of SF_6 puffer breaker and development of 300 kV–50 kA one-break circuit breaker'. IEEE PES Summer Meeting, Portland, paper 81 SM 470-4, pp. 1–8

YANABU, S., NISHIWAKI, S., MIZOGUCHI, H., SHIMOKAWARA, N., and MURAYAMA, Y. (1982): 'High current interruption by SF_6 disconnecting switches in gas insulated switchgear', *IEEE Trans.,* **PAS-101,** pp. 1105–1113

YANABU., S., TAMAGAWA, T., IROKAWA, S., HORIUCHI, T., and TOMINURO, S. (1981): 'Development of HVDC circuit breaker and its interrupting test'. IEEE PES Summer Meeting, Portland, paper 81 SM 473-8, pp. 1-8

Yaskawa Elect. Mfg. Co., Ltd. Japan: Fluopac Series Rotary arc circuit breakers

YOSHIOKA, Y., and NAKAGAWA, Y. (1978): 'Investigation of the interrupting performance of a puffer type gas circuit breaker'. IEEE PES Summer Meeting, Los Angeles, paper A 78 597, pp. 1-7

YOSHIOKA, Y., and NAKAGAWA, Y. (1980): 'Investigation of interrupting performance of puffer type gas blast circuit breaker under various nozzle clogging conditions'. IEEE PES Winter Meeting, New York, paper F 80, pp. 281-286

YOSHIOKA, Y., TSUKUSHI, M., and NATSUI, K. (1979): 'A method and application of a theoretical calculation for on-load pressure rises in puffer type gas circuit breakers', *IEEE Trans.*, **PAS-98**, pp. 731-737

YOSHIZUMI, Y. *et al.* (1982): 'Fast transient overvoltages in GIS caused by the operation of isolators'. Proc. 3rd Int. Symp. on Gaseous dielectrics, p. 456

ZHANG, J., XU., G. and XU, C. (1985): 'Research into rotating arc in an SF_6 circuit breaker'. Proc. Int. Conf. on Gas discharges and their applications, Oxford (Leeds University Press)

ZIMMERMANN, H. (1984): 'The self-extinguishing principle using SF_6: an economical development in circuit breaker design'. Proc. IEE Symp. on Trends in modern switchgear design 3·3–150 kV, Newcastle, pp. 4.1–4.4

Index

Ablation 33, 34, 37, 72, 174
  layer 37
  products 48
AC
  asynchronous 107
  component 106
  links 107
Aerodynamics 63
Air 1, 3, 6, 12, 15, 30, 77, 80, 81, 84, 107,
    116, 118, 128, 139,162, 165, 166, 167
  blast 53, 80, 106, 165
  compressed 84
  gaps 98
  plasma 11
Alazarin complexon 163
Alumina 164, 176
Aluminium 94, 100, 101, 102, 121, 126, 141,
    164, 177
Analogue,
  electrical 64
ANSI 143, 147
Arc 1, 13, 34, 122, 173
  axisymmetric 13, 37
  by products 80, 103, 121, 123, 176
  chamber 28, 32, 52, 53, 72
  characteristic 111
  column 22, 37, 54, 55, 58, 65, 69, 73, 167,
    169, 170, 173
  contact 73
  control 6, 7, 11, 18, 37, 42, 62, 63, 122,
    172
  crossflow 13, 37
  cross section 50, 173
  discharge 64, 164, 173, 175
  duration 101
  energy 101, 102, 149, 152
  extinction 98, 120, 128, 134

  free burning 17, 132
  furnace 116, 118, 120
  gas blast 43, 170
  heating 13, 23, 27, 60, 72, 99, 100, 137,
    140
  helical 38, 122, 123
  high current 7, 34, 37, 60, 73
  instability 22, 91, 117
  lengthening 17
  modelling 4, 51, 63, 65, 69
  motion 101
  plasma 13, 15, 37, 107, 170, 173
  pressurisation 17, 53
  properties 170
  quenching 6, 15, 17, 18, 37, 52, 53, 62, 63,
    77, 116, 121, 138,170, 172
  recovery 23, 28, 30, 71
  reignition 42, 118
  resistance 106, 107, 111, 112, 148
  root 37, 38, 100, 107
  rotation 20, 22, 23, 30, 33, 73, 93, 122,
    124, 136
  rupture 174
  stability 42, 43
  switchgear 13
  time constant 111, 112
  transfer 122
  velocity 75
  voltages 75, 101, 111, 122, 152, 160
Area
  characteristic 70
  conductance 70
  flow 72
  thermal 71, 72
Arrester 111, 113, 160
  surge 95, 97, 98, 99, 157, 158, 168, 169
Auto-reclose 1, 98, 117, 131, 155, 178

Index

Barium chloride 176
Bentonite thickener 165
Brass 164
Boiling point 62
Boundary layer analysis 70, 72
Breakdown 143
  dielectric 52, 55, 73
  electrical 89
  gas-gap 177
  insulation 95
  strength 58, 61, 62
  surface 61
Breaking capacity 81
Breaks per phase 95
Browne model 69
Burn through 100, 101, 103
Busbar
  duplicate 89
  gas insulated 84
  isolator 114
  systems 88, 89, 115, 166
Bus schemes 88
Bushing
  air/SF_6 94
  epoxy resin 121
  gas filled 83
  transformer 131

Cable
  insulated 84
  oil filled 114
  termination 88
CAD 4, 62, 63, 179
Capacitor
  back-to-back 143
  divider 168, 169
  grading 65, 84
  power factor 118
  surge 128
Carbon 45, 61
Cassie model 69, 72
Cathode 61
CEGB 114
Ceramic 163
CF_4 14, 36, 52, 163
C_2F_6 52
CH_4 52
Charge
  space 56
CIGRE 63, 136, 163
Circuit
  coupled 64, 65
  instability 1

interrupter 128
interruption 1, 103
  short 106, 117, 125, 128, 131, 148, 155, 175
  synthetic 147, 149, 154, 157, 165
Circuit breaker 77, 88, 94, 98, 99, 104, 106,
  107, 108, 110, 111,112, 113, 116, 117,
  118, 120, 121, 124, 125, 127, 128, 131,
  134,135, 137, 138, 139, 142, 143, 144,
  146, 147, 149, 151, 154, 155,158, 160,
  163, 164, 167, 168, 174, 175, 176, 178,
  179
  air blast 80, 106, 135, 139, 165
  auxilliary 144, 149, 154, 160
  bulk oil 80
  D.C. 107, 108, 109, 158, 160
  dead tank 77, 82, 83, 84, 88
  duo flow 107, 138, 176
  EHV 60
  failures 136, 167
  feeder 131
  gas blast 170
  generator 80, 104, 106, 107, 125, 165
  high capacity 73
  high voltage 51
  live tank 82
  low capacity 73
  low oil 3, 81
  minimum oil 80, 81, 135, 165, 167
  multi break 151, 152, 168
  multi terminal 108
  oil 123, 138, 166, 167, 179
  open terminal 78
  parallel 166
  partial duo flow 138
  puffer 65, 77, 81, 107, 120, 121, 124, 125,
    127, 134, 138, 140,142, 156, 166, 172
  rotary arc 122, 124, 125, 174
  self pressurising 124, 125, 135
  SF_6 77, 81, 102, 107, 121, 130, 139, 152,
    165, 167
  single pressure 80, 107, 139
  two pressure 77, 135
  vacuum 3, 120, 130, 167
CO_2 162, 163
Compression 72
Computer packages 63
Conductance 70, 75
Conductivity
  electrical 12, 13, 14, 33, 70
  thermal 7, 33, 34, 120
Conservation
  energy 65, 69

equations 65, 69
mass 65, 69
monumentum 65, 69
Contact 30, 43, 44, 60, 61, 65, 73, 118, 121, 127, 166, 167, 174
  annular 20, 28, 44
  cylindrical 53
  diameter 28
  drive 63
  fixed 65
  erosion 107
  material 44, 53, 120, 174
  metallic 34
  movement 25, 55, 138
  resistance 106
  ring 122
  separation 15, 16, 25, 27, 28, 31, 42, 65, 107, 122, 134, 138
  speed 123
  stroke 60
  travel 27, 60, 81, 133, 138
  upstream 37, 65, 173, 178
  vapour 13, 15, 17, 34, 48
  wear 103, 166
  welding 154
  worn 162
Contamination 13, 107, 126, 162
  particle 88
Convection 30, 75
  axial 69
  cross flow 30
  forced 17, 106
Copper 45, 141, 163, 175
  tungsten 34
  vapour 7, 13
Corrosion 94, 113, 127
Crank Nicholson 65
Crossflow 20, 37
CS 52
Current
  arcing 60, 140, 152, 175
  assymmetry 42, 103
  capacitive 89, 144, 154
  chopping 117, 118, 120, 122, 128, 144, 165
  closed loop 91, 156
  decay 48, 54
  density 37, 44, 51, 71, 167, 170
  eddy 42
  fault 3, 16, 17, 20, 25, 34, 37, 42, 44, 60, 74, 80, 100, 101,104, 106, 110, 112, 120, 124, 128, 130, 135, 138, 155, 166, 179
  inductive 98, 99, 125, 143, 154, 165

injection 148, 151
inrush 165
interruption 42, 53, 61, 89, 103, 132, 143, 152, 154, 156, 160, 164
magnetising 117
peak 23, 27, 28, 43, 50, 140, 160, 167, 175
post arc 75, 167
short circuit 3, 60, 103, 106, 143, 155, 166
transformer 77, 82, 88, 94, 168
waveform 27, 42, 44, 138, 151
zero 1, 15, 17, 27, 43, 48, 49, 50, 55, 56, 65, 69, 75, 103, 104,106, 107, 108, 110, 111, 117
Cytotoxic 165

D_2 52
D.C. 104
  components 106
  converters 112
  interruption 54
  interrupter 112
  transmission 3, 80, 107, 179
Decompression 23, 28, 30, 51, 52
Density
  electron 175
  gas 67
  magnetic flux 20, 75
  thermal 71
Design
  computer-aided 4, 63
  evaluation 63
Dielectric
  breakdown 52, 55, 73
  failure 61, 88
  integrity 98
  properties 112, 166
  recovery 6, 7, 50, 54, 56, 58, 59, 60, 61, 68, 143, 166
  strength 6, 56, 62, 80, 94, 120, 125, 126, 127, 132
  stress 60, 83, 89, 151
Diffusion
  radial 7, 69
Discharge
  glow 111, 112
  partial 167, 176
  surface 127
Disconnector 88, 89, 92, 94, 99, 132, 149
Dissociation 12, 14
  fragments 7
Distribution
  primary 116, 117, 130
  secondary 116, 127, 128

switchgear 116, 118, 126
  voltage 126, 127
Drag 13, 42, 73
Drive
  contact 63
  electromagnetic 141
  energy 18, 123
  gear 138
  magnetic 133, 140
  mechanical 63, 120, 124, 130, 137, 140, 170
  mechanism 120, 160, 161

Echinghen converter station 112
Elastomeric books 130
Elektrizitats-Gessellschoft Laufenbury AC 114
Electrode 164
  material 71, 163, 174
  shield 169
  vapour 48, 75, 176, 178
Electrons 44, 174
Electrostatic
  shield 61, 168
Emission
  coefficient 7, 13
  radiative 75
Energy
  absorber 111, 158, 160
  compression 135
  drive 18, 21, 135
  mechanical 139
  operating 137
  spring 139
  thermal 12, 102
Enthalpy 8, 12
  removal 12
  specific 12
  stagnation 71
Equilibrium
  non-thermal 69
  thermal 65
  thermodynamics 58
Exothermic 102
Expansion
  thermal 135
Evaporation
  contact 44

Failure
  dielectric 61, 88
  mechanical 136, 167
  thermal 61

Fault 100
  arc 162, 170
  current 3, 16, 17, 20, 25, 34, 37, 42, 44, 60, 74, 80, 100, 101,104, 106, 112, 120, 124, 128, 130, 135, 138, 155, 166, 179
  detection 131
  earthing 176
  evolving 143, 165
  ground 157
  interphase 95
  interruption 89, 104, 143, 148, 149
  inverter station 108
  mechanical 103
  operation 118, 161
  phase-to-ground 155
  protection 104
  short circuit 148
  short line 139, 143, 147, 149, 154, 165
  statistics 101
  terminal 73, 108, 139, 143, 154
Field
  aerodynamic 63, 173
  calculations 64
  critical strength 61
  distribution 65
  electromagnetic 63, 64, 65
  electrostatic 63, 65
  flow 64, 65, 133, 173
  magnetic 38, 42, 43, 65, 73, 122, 133
  thermal 63, 173
Finite
  difference 63, 68
  element 64
Flashover 89, 95, 98, 99, 121, 139, 162, 164, 177
Flow
  air 139
  area 72
  axial 64, 73
  blocking 15, 44
  cross- 30, 37, 73
  crossection 22, 51, 71
  duo- 45, 48, 59, 113, 176
  entrainment 65
  gas 1, 122, 173
  inertia 23
  incompressible 64
  instabilities 64
  mass 81
  mono- 44, 45, 175
  nozzle 122
  reverse 23, 34

196 Index

sonic 44, 49
straining 70
subsonic 50
supersonic 15
throttling 44, 137
velocity 48
vortex 65
Fluorides 163, 176
Fluorine 13, 56
Flux
　magnetic 20, 38, 42, 43, 53, 54, 75
Force
　electromagnetic 42, 65, 100, 123
　magnetic 37, 42, 124
　tripping 139
Freon 84
Fuse 127, 128, 131, 167
　high voltage 128, 130
　switch 128, 131

Gaps
　air 98
　arc 69
　coordinating 98
　gas 162, 164, 177
　rod 98
　trigger 111, 144
Gas
　analysis 163
　compressed 15, 31, 134, 135, 137, 166
　chromatography 163, 176
　compressed 15, 31, 134, 135, 137, 466
　filled enclosure 102
　flow 1, 11, 122, 173
　heating 75, 124
　insulated 85, 91, 99, 100, 112, 113, 115, 169
　ionised 88
　law 23, 72
　leakage 118, 121
　mixing 72
　pressure 1, 101, 106, 118, 122, 124, 137
GIS 85, 87, 91, 95, 98, 99, 100, 126, 127, 143, 160, 163, 169, 170
　hybrid 85, 87
Generator
　circuit breaker 80, 104, 106, 107
　transformer 103
Glass 163
Graphite 13, 44
　nozzle 107

H_2 52

H_2O 163
HCl 176, 177
H_e 52, 163
Heat capacity 13
HF 6, 163, 165, 176, 177
Humidity 125, 127
HVDC 104, 110

IEC
　standards 5, 61, 95, 110, 143, 147, 163
Impedance
　surge 95, 117, 118, 148
Impurities 162
　absorbants 103
　particles 94
Installation
　hybrid 114
　metal clad 84
Insulated system
　conventional 84, 87
　gas 84, 87, 99
　hybrid 84, 87
　metal clad 87
Insulation
　air 114, 116, 127, 130
　breakdown 95
　cast resin 127
　coordination 89, 95, 99, 114
　degradation 166
　electrical 1
　high voltage 89, 113, 114
　integrity 171
　level 95
　oil 116, 118
　phase-to-phase 89
　porcelain 98
　SF_6 95, 114, 115, 127
　solid 103, 127
　strength 2, 68, 93, 126
　weakness 65
Integral analysis 71
Interrupter
　air 118, 128
　chamber 54, 84
　D.C. 1, 54
　dead tank 77, 84
　duo flow 16, 17, 35, 37, 65, 72, 73, 77, 137
　EHV 15, 27, 31, 33, 60, 134
　free burning 132, 133
　hybrid 20, 30, 33
　monoflow 15, 22, 35, 93
　oil 118

Index 197

partial duoflow 15, 22, 31, 34, 48, 65, 72, 137
puffer 1, 13, 15, 16, 20, 22, 23, 27, 28, 30, 33, 36, 43, 44, 50,52, 54, 61, 68, 72, 77, 80, 93, 112, 120, 121, 130, 131, 132,134, 137
rotary arc 1, 13, 20, 21, 22, 30, 33, 38, 43, 53, 54, 61, 65, 69,73, 120, 123, 128, 131, 132, 140, 178
self-pressurising 1, 13, 20, 28, 81, 93, 107, 120, 132, 133, 137,178
SF_6 22, 44, 80, 93, 112, 117, 120, 128, 133, 147, 149,160, 178, 179
single pressure 81
suction 1, 28, 132, 133, 137, 178
two pressure 15, 22, 23, 28, 43, 44, 50, 52, 53, 80, 81, 134
units 54, 120, 167
vacuum 80, 112, 118, 120, 127, 128, 131, 178, 179
Interrupting
  ability 51, 120, 125, 168, 174
  capacity 81, 137
Interruption
  charging current 157
  fault 89, 143
  parallel 143
Ionic recombination 61
Ionisation potential 62
Ions
  negative 7, 56
  recombination 44
Isolators 88, 111
Itaipu hydro power plant 114

Jet
  electrode 174
  vapour 174
JISC 163

Leakage 126
  gas 118, 121
  oil 136
Lethal dose low 164
Lighting 131
  impulse 89
  overvoltage 98
  strokes 99
  surges 89, 98, 158
Lines
  artificial 148
  overhead 94, 99, 131

Linkages
  mechanical 140, 170
Liquefaction 13, 51, 53, 78, 80, 127
Load breaking 89
Losses
  convective 73, 75
  corona 147
  electromagnetic 93
  transmission 107
Lorenz force 20, 38, 42, 73

Mach number 48, 64
Maintenance 94, 103, 117, 118, 123, 125, 168
Mass
  ablation 23, 34, 35
  density 11, 12, 22, 61
  flow 12, 28, 52, 53
  injection 72
Mayr model 50, 69, 72
Maxwell's equations 69
Mechanism
  electromagnetic 134
  explosive 134
  hydraulic 134, 136, 139, 140
  low energy 134
  operating 1, 81, 134, 136, 138, 152, 167, 170
  pneumatic 134, 139
  puffer-drive 140
  spring 131, 134, 139, 142
Metal clad 84, 88, 94, 95, 98, 99, 100, 103, 113, 114, 132, 140
  arcs 100
  rupture 100
Metal oxide 98, 99, 128
Microprocessor control 131, 171
Modulator system 83
Momentum
  transfer 13, 73
Moisture 162, 163
Monitoring
  in service 169

N_2 30, 52, 78, 79, 127
Nitric acid 127
Nitrogen oxides 127
None equilibrium 13
Nozzle 11, 15, 16, 22, 27, 30, 37, 43, 44, 51, 53, 60, 61, 65,73, 106, 173
  ablation 22, 33, 34, 36, 37, 71, 72, 165
  area 72, 106
  axis 64

198 Index

blocking 17, 22, 33, 44, 71, 72, 81, 137, 138, 140, 170, 173
coefficient 71
diameter 37
duo flow 34, 45, 46, 48, 53, 55, 59, 64
flow 122
geometry 27, 36, 44, 48, 51, 58, 64
graphite 107
insulating 15, 35
material 45, 60, 71, 174
metallic 34, 35, 37, 55
monoflow 36, 44, 45, 46, 48
PTFE 36, 61, 163
shape 60
throat 27, 28, 31, 34, 36, 48, 51, 55, 60, 61, 72, 113, 175
transparent 174
wall 22
wear 103

Ohm's law 69, 80, 84
Oil 1, 3, 6, 116, 117
  actuator 140
  leakage 136
  mineral 163, 165
  vacuum 173
Open
  layout 87
  terminal 82, 84
Operation
  close-open 125
  closing 142
  one cycle 138
  spring 127
Optical fibres 168, 169, 171, 173, 175, 179
Organic greases 163
Outage 117, 155, 178
Over current 130, 131
Over pressurisation 100, 102, 139
Over voltage 54, 95, 97, 98, 99, 117, 120, 122, 125, 128, 138,143, 169
  arc extinction 98
  atmospheric 97
  lightning 97, 98
  suppression 125
  switching 98, 99
  system 97, 103
  temporary 98
  transient 97
Ozone 127

Pacific Intertie 112, 113

Particle
  density 59
  impurities 94
  trap 94
Paschen's law 61
Performance
  interruption 48
Phase
  angle 53
  lag 42, 43, 53
  segregation 84, 121
Photography 174
Piston 137, 138
  auxilliary 20, 124
  chamber 27, 30, 32, 36, 52, 72, 121, 138, 140, 142, 173
  compression 15, 17, 25, 27, 28, 50, 72
  drive 17, 27, 28
  electromagnetic 17
  speed 72
  stroke 27, 51
  suction 18
  travel 50
Plasma 170, 173
  air 11
  compression 37
  copper 11
  flow 37, 173
  properties 174
  temperature 50
Polarisability 62
Pollution 113, 114
Potential
  dividers 168
  ionisation 62
  magnetic vector 65
Power
  loss, dynamic 71
  radiated 7, 34
  transformer 99
Prearcing 154, 155
Pressure
  arc-induced 27, 28, 53, 54, 103, 133
  background 50, 54, 133
  elevation 22, 23, 27, 30, 31, 33, 36, 72, 101, 102
  gas 1, 53, 101, 106, 118, 122, 137
  gradient 22
  increases 12, 50
  measurement 34
  nominal 53
  oscillations 22
  puffer 27, 51

Index 199

relief 102, 103
self pressurisation 93
transient 22, 27, 28, 30, 32, 37, 81, 113
Probe
 inductive 170
 radio frequency 170
 voltage 170
PTFE 34, 45, 67
 ablation 34
 dissociation 13
 glass filled 45
 nozzle 34, 103
Pumped storage 93, 106, 107, 114

Quality control 95, 121
Quartz 163

Radiation
 flux 34
 loss 13, 72, 74
 optical 22, 173, 175
 transfer 7, 34, 69
 transport 69
Reactors
 smoothing 110
Recombination 56
 ionic 61
Recorder
 travel 170
Recovery
 arc 71
 dielectric 6, 7, 50, 54, 56, 58, 59, 60, 61, 65, 68, 143, 166
 electrical 124
 free 75
 period 15, 27, 28, 30, 32
 thermal 7, 43, 44, 50, 52, 60, 64, 71, 73, 75, 140
 voltage 48, 54, 56, 67, 73, 91, 123, 132, 143, 147, 148, 149,160, 165
Reignition
 critical 75
 thermal 53
Reliability 82, 88, 95, 103, 104, 108, 113, 114, 134, 136, 144, 167
Resistor
 closing 82, 84, 138
 current limiting 99
 metal oxide 128
Restrike 122, 144, 147, 148
 surge 99
Resonance
 aerodynamic 22, 23, 30

electron paramagnetic 163
 nuclear magnetic 163
Ring bus 165
Ring main unit 128, 130, 131
Rod gaps 98
RRRV 126, 157
Rollarc contactor 122

Schlieren technique 173
Safety 95, 100
Scaling 37, 48, 58
 laws 36, 37, 58, 70
 parameter 33
Sealed for life 118, 123, 179
Sealing agents 163
Self extinguishing principles 80
SF_4 163, 176
SF_6 6, 13, 15, 28, 43, 52, 53, 61, 62, 77, 78, 79, 80,84, 94, 98, 99, 102, 103, 106, 107, 113, 114, 116, 117, 118, 127,128, 132, 135, 162, 164, 165, 166, 167, 175, 178
 Cu 13
 byproducts 94, 162, 163, 165, 176, 177
 dissociation 6, 7, 12
 hypercritical 112
 insulation 95, 98, 113, 114, 115, 126
 leaks 94
 liquefaction 53, 80, 139
 liquid 53
 mixtures 14, 43, 51, 52, 54, 127
 N_2 13, 28, 30, 32, 52, 54, 79
 properties 11, 12
 PTFE 13
 sparked 165
 switchgear 114, 120
S_2F_{10} 164
Shape factors 71
Short circuit tests 83
Shunt
 resistive 167
Si 164
SiC 98, 99
SiF_4 163, 164, 165
SiO_2 163
Silica 163
Silicone rubber 163
Society of Electrical Cooperative Research 163
SO_2 163, 164, 165, 176, 177
SOF_2 163, 165
SO_2F_2 163, 164, 165
SOF_4 164, 165

Specific heat 30, 72
  volume 11, 12
Spectronomy
  mass 163
Spectroscopic 34, 163, 173, 174, 175
  point 56
Stagnation region 45, 56, 59, 60, 65
Standards
  IEC 5, 17
Steel 45, 94, 101, 126, 164
Substation
  air insulated 98, 114
  gas insulated 100
  high voltage 88
  layout 84, 95, 179
  metal clad 2, 84, 98
Suction chamber 28, 72, 139
Sulphur 56, 165, 177
Surge
  arrester 95, 97, 98, 99, 111, 157, 158, 168, 169
  capacitor 128
  impedance 95, 117, 118, 148, 149
  lightning 89, 158
  restrike 98, 99, 157
  voltage 157
Swirl 22, 37
Switch
  earthing 88, 91, 92, 93, 116, 132
  disconnect 89, 132, 149, 156, 157, 169
  load breaking 128, 131
  vacuum 113
Switchgear 117
  distribution 116, 118, 126
  SF_6 114, 120
  transmission 3
Switching
  capacitive 143, 165
  duty 91
  out of phase 143, 154
  surge 144
  time 112

Testing
  circuit breaker 144, 147, 160, 167
  development 167, 175
  direct 144
  full pole 151, 152, 167
  full power 167
  performance 178
  short circuit 146, 167, 170, 175
  synthetic 146, 147, 149, 157, 165, 167, 179
  type 116, 117

unit 148
Tests
  bias 144
  closing 154
  dielectric 144
  synthetic 3, 157, 179
Thermal
  area 71, 72
  capacity 11
  conduction 7, 75, 120
  diffusivity 34, 100
  failure 61
  recovery 55, 58, 140
  reignition 53
  stress 89
  volume 22, 72, 173
Threshold limit value 164
Time
  acoustic transit 173
  operating 124
Time constant 110, 112
  arc 133, 140, 160
Toxic
  agents 164
  fragments 6
Toxicity 62, 121, 162, 164, 165
Three phase 126, 127
  enclosure 95, 149
Transducers 172
Transformer
  current 77, 94, 168
  explosions 139
  power 99
  switching 117
  tee switch 128, 130
  voltage 88, 94
Transmission
  ENV 3
  DC 3
  UHV 3
Trefoil geometry 124
Turbulence 13, 22, 58, 65, 69, 70, 173

Utilities 118
Urea thickener 165

Vacuum 116, 117, 118, 128, 160
Vapour
  metal 16
Velocimetry
  laser-Doppler 173
Velocity
  axial 101

rotational 42
sound 12, 51, 52, 65, 72
Viscosity 13
Voltage
  arc 75, 122, 152
  breakdown 58, 60, 61, 143
  contact-contact 158
  distribution 126, 127, 154, 170
  divider 168
  injection 149
  phase to
    ground 158
    phase 149, 154, 158
    tank 152, 154
  probes 170
  recovery 48, 49, 50, 54, 56, 60, 73, 91, 123, 132, 143, 147, 148,149, 160, 165
  restrike 65, 148, 157
  sharing 167
  surge 157

transformer 88, 94, 131
transient 56, 60, 99, 146, 147, 158, 169
transmission 125
withstand 61, 89, 111, 112, 126
Vortex 13, 37, 65

Wake 42, 65
Wave
  point on 25, 31, 42
  shock 165, 173
Waveform
  current 151, 173
  EXP-COS 147
  restrike 148
  two parameter 148
  voltage 147
Weil circuit 147, 148, 149, 155
Weight/capacity ratio 81

Z_nO 98, 99, 111, 113, 160